Advanced Maths Essentials
Core 1 for OCR

Welcome to Advanced Maths Essentials: Core 1 for OCR. This book will help you to improve your examination performance by focusing on all the essential maths skills you will need in your OCR Core 1 examination. It has been divided by chapter into the main topics that need to be studied. Each chapter has then been divided by sub-headings, and the description below each sub-heading gives the Edexcel specification for that aspect of the topic.

The book contains scores of worked examples, each with clearly set-out steps to help solve the problem. You can then apply the steps to solve the Skills Check questions in the book and past exam questions at the end of each chapter. If you feel you need extra practice on any topic, you can try the Skills Check Extra exercises on the accompanying CD-ROM. At the back of this book there is a sample exam-style paper to help you test yourself before the big day.

Some of the questions in the book have a (disc) symbol next to them. These question PowerPoint® solution (on the CD-ROM) that guides you through suggested steps i the problem and setting out your answer clearly.

Using the CD-ROM

To use the accompanying CD-ROM simply put the disc in your CD-ROM drive, and the menu should appear automatically. If it doesn't automatically run on your PC:

1. Select the My Computer icon on your desktop.
2. Select the CD-ROM drive icon.
3. Select Open.
4. Select core1_for _ocr.exe.

If you don't have PowerPoint® on your computer you can download PowerPoint 2003 Viewer®. This will allow you to view and print the presentations. Download the viewer fro http://www.microsoft.com

Pearson Education Limited
Edinburgh Gate
Harlow
Essex
CM20 2JE
England
www.longman.co.uk

First published 2005
Second impression 2008
ISBN   978 0 582 836495

Design by Ken Vail Graphic Design

Cover design by Raven Design

Typeset by Tech-Set, Gateshead

Printed in the U.K. by Scotprint, Haddington

The publisher's policy is to use paper manufactured from sustainable forests.

The publisher wishes to draw attention to the Single-User Licence Agreement at the back of the book. Please read this agreement carefully before installing and using the CD-ROM.

We are grateful for permission from the OCR to reproduce past exam questions. All such questions have a reference in the margin. OCR can accept no responsibility whatsoever for accuracy of any solutions or answers to these questions.

Every effort has been made to ensure that the structure and level of sample question papers matches the current specification requirements and that solutions are accurate. However, the publisher can accept no responsibility whatsoever for accuracy of any solutions or answers to these questions. Any such solutions or answers may not necessarily constitute all possible solutions.

1 Indices and surds

1.1 Indices

Understand rational indices (positive, negative and zero) and use laws of indices in the course of algebraic applications

The laws or rules of indices allow you to simplify terms that are written in **index form**, a^m, where m is rational.

a is the **base**, where $a \neq 0$.

m is the **index**, also known as the **power** or **exponent**.

These rules apply to terms in index form with the same base:

> ***Rule 1*** To **multiply** the terms, **add** the indices:
> $$a^m \times a^n = a^{m+n}$$
>
> ***Rule 2*** To **divide** the terms, **subtract** the indices:
> $$a^m \div a^n = a^{m-n}$$
>
> ***Rule 3*** To **raise to a power**, **multiply** the indices:
> $$(a^m)^n = a^{mn}$$

Note:
The index m can be a positive or negative integer, or a fraction, or zero.

Tip:
$a^m \div a^n$ is also written $\dfrac{a^m}{a^n}$.

Example 1.1 Simplify **a** $2x^5 \times 7x^6$ **b** $10x^5z^3 \div 2x^3z$ **c** $(2p^2)^5$

Step 1: Gather like terms and simplify using the index laws.

a $2x^5 \times 7x^6 = 2 \times 7 \times x^5 \times x^6$
$$= 14 \times x^{5+6} \qquad\qquad (Rule\ 1)$$
$$= 14x^{11}$$

b $10x^5z^3 \div 2x^3z = \dfrac{10x^5z^3}{2x^3z}$
$$= 5x^{5-3}z^{3-1} \qquad\qquad (Rule\ 2)$$
$$= 5x^2z^2$$

c $(2p^2)^5 = 2^5(p^2)^5 = 32p^{10} \qquad\qquad (Rule\ 3)$

Tip:
Multiply or divide the numbers, then deal with the expressions in index form.

Tip:
When no index is written, this means the power is 1, so $z = z^1$.

Tip:
$(ab)^n = a^nb^n$, so remember to raise 2 to the power 5 here.

Example 1.2 Write each of these expressions as a power of 2:

a 8^4 **b** 4^{x+1}

Step 1: Write each term in index form with the same base and simplify using the index laws.

a $8^4 = (2^3)^4 = 2^{12} \qquad\qquad (Rule\ 3)$

b $4^{x+1} = (2^2)^{x+1} = 2^{2(x+1)} = 2^{2x+2} \qquad\qquad (Rule\ 3)$

The zero index, a^0

You know that $a^n \div a^n = 1$

But, by *Rule 3*, $a^n \div a^n = a^{n-n} = a^0$

$$\Rightarrow \qquad\qquad a^0 = 1 \qquad\qquad \textbf{\textit{Rule 4}}$$

Note:
a cannot be zero; 0^0 is undefined.

Negative index, a^{-n}

You know that $\qquad a^n \times a^{-n} = a^0 = 1$

Divide both sides by a^n $\qquad a^{-n} = \dfrac{1}{a^n}$ $\qquad\qquad$ ***Rule 5a***

This format is useful when working in fractions:

$$\left(\frac{a}{b}\right)^{-n} = \left(\frac{b}{a}\right)^{n}$$ $\qquad\qquad$ ***Rule 5b***

Fractional indices

$$a^{\frac{1}{n}} = \sqrt[n]{a}$$ $\qquad\qquad$ ***Rule 6a***

For example, $a^{\frac{1}{3}} = \sqrt[3]{a}$.

$$a^{\frac{m}{n}} = (\sqrt[n]{a})^m = \sqrt[n]{a^m}$$ $\qquad\qquad$ ***Rule 6b***

For example, $a^{\frac{2}{3}} = (\sqrt[3]{a})^2 = \sqrt[3]{(a^2)}$.

To calculate $64^{\frac{2}{3}}$ you could find $(\sqrt[3]{64})^2 = 4^2 = 16$.

Alternatively, you could find $\sqrt[3]{(64^2)} = \sqrt[3]{4096} = 16$.

Example 1.3 Evaluate, without using a calculator,

a $3^4 \div 3^7$ $\qquad$ **b** $\left(\frac{3}{4}\right)^{-1}$ $\qquad$ **c** $4^{\frac{1}{2}}$ $\qquad$ **d** $8^{-\frac{1}{3}}$ $\qquad$ **e** $\left(\frac{1}{125}\right)^{-\frac{2}{3}}$

Step 1: Use the index laws to calculate the values.

a $3^4 \div 3^7 = 3^{-3} = \dfrac{1}{3^3} = \dfrac{1}{27}$ $\qquad\qquad$ *(Rules 2 & 5a)*

b $\left(\frac{3}{4}\right)^{-1} = \left(\frac{4}{3}\right)^1 = \frac{4}{3}$ $\qquad\qquad$ *(Rule 5b)*

c $4^{\frac{1}{2}} = \sqrt{4} = 2$ $\qquad\qquad$ *(Rule 6a)*

d $8^{-\frac{1}{3}} = \dfrac{1}{8^{\frac{1}{3}}} = \dfrac{1}{\sqrt[3]{8}} = \dfrac{1}{2}$ $\qquad\qquad$ *(Rules 5a & 6a)*

e $\left(\frac{1}{125}\right)^{-\frac{2}{3}} = 125^{\frac{2}{3}} = (\sqrt[3]{125})^2 = 5^2 = 25$ $\qquad\qquad$ *(Rules 5b & 6b)*

SKILLS CHECK **1A: Indices**

1 Simplify:

a $2x^3y^5 \times 3xy^{-1}$ $\qquad$ **b** $(2a^2)^4$ $\qquad\qquad$ **c** $14pq^7 \div 2p^2q^5$

2 Evaluate, without using a calculator:

a $\dfrac{1}{2^{-1}}$ $\qquad\qquad$ **b** 3^{-2} $\qquad\qquad$ **c** $27^{-\frac{1}{3}}$

d $16^{\frac{3}{2}}$ $\qquad\qquad$ **e** $0.25^{-\frac{1}{2}}$

3 Simplify $3a^2b^{-2} \times 4a^3\sqrt{b}$.

4 Write as a single power of x:

   a $x^2\sqrt{x}$ **b** $\dfrac{\sqrt{x}(\sqrt{x})^3}{x^3}$ **c** $\dfrac{\sqrt{x}(\sqrt{x})^3}{x^{-3}}$

5 Write $\dfrac{p^{\frac{1}{6}}p^{\frac{2}{3}}}{\sqrt{p}}$ in the form p^k where k is a number to be found.

6 Write each of the following as a power of 2:

   a 4^x **b** 8^{x-1}

7 Write down the exact value of

   a $(2^3 + 3^2 - 1^3)^{\frac{1}{2}}$ **b** $(1^3 + 4^{\frac{1}{2}})^3$

8 Write down the exact value of

   a $[2(9^{\frac{1}{2}} + 9^2) + 1^{\frac{1}{3}}]^{\frac{1}{2}}$ **b** $(3^2 - 2^3 - 1^3)^{\frac{1}{2}}$

SKILLS CHECK **1A EXTRA** is on the CD

1.2 Surds

Recognise the equivalence of surd and index notation. Use simple properties of surds, such as $\sqrt{12} = 2\sqrt{3}$, including rationalising denominators of the form $a + \sqrt{b}$.

Some square roots cannot be written as a fraction or as a terminating or recurring decimal. They are irrational and can be written as **surds**.

Examples of surds are $\sqrt{2}$, $\sqrt{3}$, $\sqrt{5}$.

Like surds can be combined by **adding** or **subtracting**:

$$5\sqrt{3} + 4\sqrt{3} + 6\sqrt{5} - 2\sqrt{5} = 9\sqrt{3} + 4\sqrt{5}$$

When **multiplying**, remember that $\sqrt{p} \times \sqrt{q} = \sqrt{p \times q}$.

$$\sqrt{2} \times \sqrt{3} = \sqrt{2 \times 3} = \sqrt{6}$$

When **squaring**, use the special case $\sqrt{p} \times \sqrt{p} = \sqrt{p^2} = p$.

$$(3\sqrt{5})^2 = 3\sqrt{5} \times 3\sqrt{5} = 9 \times \sqrt{5 \times 5} = 9 \times 5 = 45$$

To **simplify** a surd such as $\sqrt{12}$ or $\sqrt{180}$, write it in the form $k\sqrt{a}$, where a does not have any factors that are square numbers.

$$\sqrt{12} = \sqrt{2 \times 2 \times 3} = \sqrt{2^2 \times 3} = \sqrt{2^2} \times \sqrt{3} = 2 \times \sqrt{3} = 2\sqrt{3}$$

$$\sqrt{180} = \sqrt{2 \times 2 \times 3 \times 3 \times 5} = \sqrt{2^2 \times 3^2 \times 5} = 2 \times 3 \times \sqrt{5} = 6\sqrt{5}$$

Note:
The value given on a calculator for a surd is only an approximation.

Note:
$\sqrt{2} = 2^{\frac{1}{2}}$

Tip:
Split the number into its prime factors to help spot factors that are square numbers.

Example 1.4 **a** Write $\sqrt{8}$ in the form $k\sqrt{2}$, where k is an integer.

b Hence simplify $6\sqrt{2} + 5\sqrt{8}$.

Step 1: Simplify the surd. **a** $\sqrt{8} = \sqrt{4 \times 2} = \sqrt{4} \times \sqrt{2} = 2\sqrt{2}$

Step 2: Add like surds. **b** $6\sqrt{2} + 5\sqrt{8} = 6\sqrt{2} + 5 \times 2\sqrt{2}$
$$= 6\sqrt{2} + 10\sqrt{2}$$
$$= 16\sqrt{2}$$

Example 1.5 Expand and simplify $(2 - \sqrt{3})(5 + \sqrt{3})$.

Step 1: Expand the brackets. $(2 - \sqrt{3})(5 + \sqrt{3}) = 10 + 2\sqrt{3} - 5\sqrt{3} - (\sqrt{3})^2$

Step 2: Simplify each term. $= 10 - 3\sqrt{3} - 3$

Step 3: Add like terms. $= 7 - 3\sqrt{3}$

If you recognise the **difference of two squares** you can multiply quickly, using $(a + b)(a - b) = a^2 - b^2$. For example

$$(3 + \sqrt{2})(3 - \sqrt{2}) = 3^2 - (\sqrt{2})^2 = 9 - 2 = 7$$

To simplify fractions with a surd in the denominator, use a special technique called **rationalising the denominator**. This involves forming an equivalent fraction by multiplying both the numerator and denominator by a quantity that makes the denominator rational.

Example 1.6 Express $\dfrac{12}{\sqrt{3}}$ in the form $k\sqrt{3}$, where k is an integer.

Step 1: Rationalise the denominator. $\dfrac{12}{\sqrt{3}} = \dfrac{12 \times \sqrt{3}}{\sqrt{3} \times \sqrt{3}}$

Step 2: Simplify if possible. $= \dfrac{12\sqrt{3}}{3} = 4\sqrt{3}$

Example 1.7 Simplify $\dfrac{2\sqrt{3} + 1}{\sqrt{3} - 1}$.

Step 1: Rationalise the denominator and simplify. $\dfrac{2\sqrt{3} + 1}{\sqrt{3} - 1} = \dfrac{(2\sqrt{3} + 1)(\sqrt{3} + 1)}{(\sqrt{3} - 1)(\sqrt{3} + 1)}$

$$= \dfrac{6 + 3\sqrt{3} + 1}{3 - 1^2}$$

$$= \dfrac{7 + 3\sqrt{3}}{2}$$

Example 1.8 A rectangle $ABCD$ has area $20\sqrt{2}$ cm^2 and length $AB = 4\sqrt{5}$ cm.

Giving your answers in simplified surd form, find **a** the length of BC

b the length of DB.

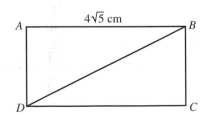

4

Step 1: Use the formula for the area of a rectangle to find the length.

a   Area $= AB \times BC$

$\Rightarrow 20\sqrt{2} = 4\sqrt{5} \times BC$

$BC = \dfrac{20\sqrt{2}}{4\sqrt{5}}$

Step 2: Rationalise the denominator.

$= \dfrac{20\sqrt{2} \times \sqrt{5}}{4\sqrt{5} \times \sqrt{5}}$

Step 3: Simplify

$= \dfrac{20\sqrt{10}}{4 \times 5}$

$= \sqrt{10}$

Length $BC = \sqrt{10}$ cm

Step 1: Use Pythagoras' theorem in triangle ADB.

b By Pythagoras' theorem

$DB^2 = AD^2 + AB^2$

$= (\sqrt{10})^2 + (4\sqrt{5})^2$

$= 10 + 16\,(5)$

$= 90$

Step 2: Simplify.

Length DB

SKILLS CHECK **1B: Surd**

1 Simplify

   a $\sqrt{50}$

   d $\sqrt{12} + 5\sqrt{3}$

2 Simplify the foll

   a $3r + 5p$

   d $(5q)^2$

   g $\dfrac{1}{r - p}$

3 Simplify

   a $(\sqrt{11} + 1)^2$ **b** $\sqrt{11} - 1$

4 Triangle ABC is right-angled at B, $AB = \dfrac{3}{\sqrt{5}}$ cm, $BC = 5\sqrt{2}$ cm. Its area is $p\sqrt{10}$ cm^2, where p is rational. Find the value of p.

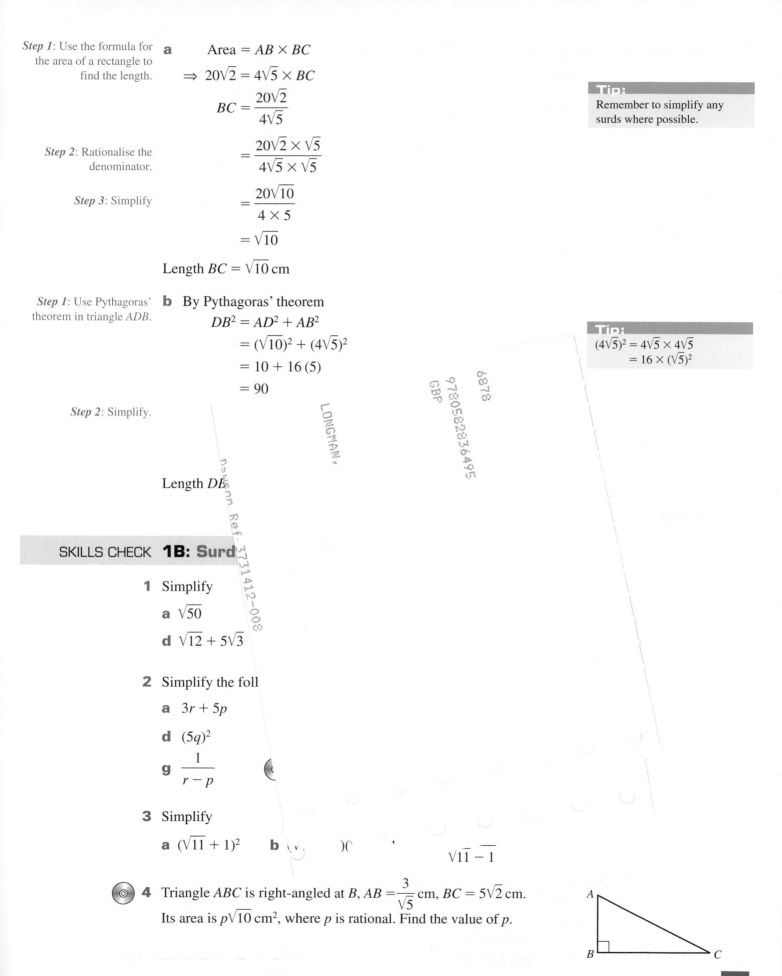

5 a Express $\sqrt{48}$ and $\dfrac{6}{\sqrt{3}}$ in the form $k\sqrt{3}$, where k is an integer.

  b Hence write $\sqrt{48} + \dfrac{6}{\sqrt{3}}$ in the form $p\sqrt{3}$, where p is an integer.

6 a Express $(\sqrt{7} + 1)^2$ in the form $a + b\sqrt{7}$, where a and b are integers.

  b Express $\dfrac{\sqrt{7} + 1}{\sqrt{7} - 1}$ in the form $c + d\sqrt{7}$, where c and d are rational numbers.

7 a Given that $(3 + \sqrt{5})(4 - \sqrt{5}) = p + q\sqrt{5}$, where p and q are integers, find p and q.

  b Given that $\dfrac{3 + \sqrt{5}}{4 + \sqrt{5}} = r + s\sqrt{5}$, where r and s are rational numbers, find r and s.

SKILLS CHECK **1B EXTRA** is on the CD

Examination practice   Indices and surds

1 i Given that $81 = 3^x$, write down the value of x.

  ii Given that $81^y = 3^{1-y}$, find the value of y.   [OCR May 2003]

2 Write down the exact values of

  i 4^{-2},

  ii $(2\sqrt{2})^2$,

  iii $(1^3 + 2^3 + 3^3)^{\frac{1}{2}}$.   [OCR Spec paper]

3 a Simplify $\sqrt{(24)} + 7\sqrt{(54)}$, giving your answer in simplified surd form.

  b Given that $\dfrac{8^n \times 2^{2n}}{4^{3n}} = 2^{kn}$, find the value of k.   [OCR Nov 2003]

4 A rectangle has sides of length $(4 - \sqrt{5})$ cm and $(3 + 2\sqrt{5})$ cm. Find the area of the rectangle, giving your answer in surd form as simply as possible.   [OCR Jan 2004]

5 Write down the exact value of $\left(\dfrac{27}{8}\right)^{-\frac{1}{3}}$.

6 Solve the equation $\sqrt{12}(x - 4) = 2x$, giving your answer in the form $a + b\sqrt{3}$, where a and b are integers to be found.

7 Express $\dfrac{4 - \sqrt{2}}{5 + \sqrt{2}}$ in the form $a + b\sqrt{2}$, where a and b are rational numbers.

8 Simplify $\dfrac{2x^4}{(2\sqrt{x})^4}$.

9 Express $(3\sqrt{5} + \sqrt{2})^2$ in the form $a + b\sqrt{c}$, where a, b and c are integers.

10 Given that $p = x^{\frac{3}{2}}$ and $q = x^{-\frac{5}{2}}$, express, in index form in terms of x:

a pq **b** $p \div q$ **c** $\sqrt{p}$

11 Simplify $(2x^{\frac{1}{2}})^4$.

12 **a** Express each of the following in the form $p\sqrt{3}$:

i $\sqrt{48}$ **ii** $\dfrac{6}{\sqrt{3}}$

b Hence write $\sqrt{48} - \dfrac{6}{\sqrt{3}}$ in the form $q\sqrt{3}$, where q is an integer.

13 **a** Given that $(5 + \sqrt{2})(3 - \sqrt{2}) = a + b\sqrt{2}$, where a and b are integers, find the value of a and the value of b.

b Given that $\dfrac{5 + \sqrt{2}}{3 + \sqrt{2}} = p + q\sqrt{2}$, where p and q are rational numbers, find the value of p and the value of q.

14 Write down the exact value of $(25^{\frac{1}{2}} + 64^{\frac{1}{3}})^{-\frac{1}{2}}$.

2 Polynomials

2.1 Manipulating polynomials

Carry out operations of addition, subtraction, and multiplication of polynomials.

A **polynomial** in x is an expression with positive integer powers of x, for example $4x^3 + 2x^2 - 7x + 6$.

The **degree** of a polynomial is the highest power of x, so this polynomial has degree 3.

When adding or subtracting polynomials, collect **like terms**.

Note:
The usual convention is to write the polynomial with the highest power of x first.

Example 2.1   Given that $f(x) = x^3 + 5x^2 + 7x - 2$ and $g(x) = 6x - 2x^3$, simplify the following: **a** $f(x) + g(x)$ **b** $g(x) - f(x)$

Step 1: Collect like terms.   **a** $f(x) + g(x) = x^3 + 5x^2 + 7x - 2 + 6x - 2x^3$

Step 2: Simplify. $= -x^3 + 5x^2 + 13x - 2$

b $g(x) - f(x) = 6x - 2x^3 - (x^3 + 5x^2 + 7x - 2)$

$= 6x - 2x^3 - x^3 - 5x^2 - 7x + 2$

$= -3x^3 - 5x^2 - x + 2$

Tip:
Like terms have identical letters and powers.

Tip:
Take care when subtracting.

You may need to multiply a polynomial by a polynomial. This is sometimes referred to as **expanding the brackets**.

Example 2.2   Expand and simplify these polynomials.

Step 1: Expand brackets.   **a** $2x(x^2 + 2x - 1) + (x - 3)(x + 2)$

Step 2: Collect like terms. $= 2x^3 + 4x^2 - 2x + x^2 + 2x - 3x - 6$

$= 2x^3 + 5x^2 - 3x - 6$

b $(x - 2)(x + 1)(x - 3)$

$= (x^2 - x - 2)(x - 3)$

$= x^3 - 3x^2 - x^2 + 3x - 2x + 6$

$= x^3 - 4x^2 + x + 6$

c $(x^2 + x - 3)(2x^2 - 4x + 2)$

$= 2x^4 - 4x^3 + 2x^2 + 2x^3 - 4x^2 + 2x - 6x^2 + 12x - 6$

$= 2x^4 - 2x^3 - 8x^2 + 14x - 6$

Tip:
Expand two of the brackets, then multiply by the third.

Tip:
Multiplying two brackets, each with three terms, will give nine terms, some of which will combine. Don't try to do this mentally.

Example 2.3   It is given that $(ax + b)(2x^2 - 3x + 5) \equiv 4x^3 + cx^2 + 7x - 5$.

Find the values of a, b and c.

Step 1: Expand the left-hand side to get the x^3-term and constant term.   Consider the x^3-terms and constant terms:

$2ax^3 + \cdots + 5b \equiv 4x^3 + cx^2 + 7x - 5$

Step 2: Equate terms to find a and b.   Equate x^3 terms: $2a = 4$ $\Rightarrow a = 2$

Equate constants: $5b = -5$ $\Rightarrow b = -1$

Step 3: Expand to get the x^2-term and solve for c.   Consider the x^2-terms:

$\cdots + 2bx^2 - 3ax^2 + \cdots \equiv 4x^3 + cx^2 + 7x - 5$

Equate x^2-terms: $2b - 3a = c$

$-2 - 6 = c \Rightarrow c = -8$

So $a = 2$, $b = -1$ and $c = -8$.

Note:
This is an identity, true for all values of x.

Note:
There is only one way to get the x^3-term and this will give the value of a. Similarly, equating the constant terms will give b.

Note:
You could have considered the x terms.

1 It is given that $f(x) = x^3 - 3x^2 + 5x + 4$, $g(x) = 2x^3 - 6x + 1$ and $h(x) = 5x^2 + 6x - 2$.

Simplify the following:

a $f(x) + 3h(x)$

b $g(x) - f(x)$

c $f(x) + h(x) - 2g(x)$

2 Expand and simplify the polynomials:

a $2x(x - 2) - (3x - 1)(x + 5)$

b $3x^2(x + 4) + (x^2 - 2x + 1)(x + 4)$

c $(x + 3)(x - 5)(x - 3)$

 d $(2x + 1)(x - 3)(x + 4)$

3 Find the values of a, b and c:

a $(ax + 3)(x - 4) \equiv 2x^2 - bx - 12$

 b $(ax^2 + bx + 3)(2x - 1) \equiv 6x^3 - cx^2 + 7x - 3$

c $(ax + 2)(bx - 1) \equiv 6x^2 + x - c$

SKILLS CHECK **2A EXTRA** is on the CD

2.2 Quadratic polynomials

The polynomial $ax^2 + bx + c$, where $a \neq 0$, is a **quadratic** polynomial.

Note:
If a is zero, there is no x^2-term.

The graph of $y = ax^2 + bx + c$ is called a **parabola**. It is a symmetrical curve with one turning point, called the **vertex**.

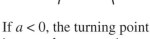

If $a > 0$, the turning point is a **minimum** turning point.

If $a < 0$, the turning point is a **maximum** turning point.

Note:
See Section 3.7 for more on finding the turning point and drawing parabolas.

The diagram shows $y = 2x^2 - 4x + 5$.

There is a minimum turning point at the vertex $P(1, 3)$.

The minimum value of y is 3.

The axis of symmetry is the line $x = 1$.

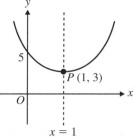

Note:
See Section 4.6 for calculus method of finding the vertex.

Find the discriminant of a quadratic polynomial.

Given a quadratic polynomial $ax^2 + bx + c$, the **discriminant** is the expression $b^2 - 4ac$.

It can be used to find

- the number of real solutions of the quadratic equation $ax^2 + bx + c = 0$ (Section 2.6)

- whether a quadratic curve crosses the x-axis at two points, touches it at one point or does not meet or cross the x-axis (Section 3.7)

- whether a line and a curve, or two curves, intersect in two places, meet at a point or do not intersect (Section 3.6).

2.4 Factorising quadratic polynomials

When you multiply two linear expressions in x, you get a quadratic polynomial, for example $(3x - 2)(4x + 5) = 12x^2 + 7x - 10$.

To **factorise** a quadratic polynomial, reverse the process and express it as the product of two linear functions.

Note:
Not all quadratic expressions can be factorised.

Some factorising can be done by guesswork, especially when the coefficient of x^2 is 1, for example:

$$x^2 + 8x = x(x + 8) \qquad \text{(common factor)}$$
$$x^2 - 9 = (x - 3)(x + 3) \quad \text{(difference of two squares)}$$
$$x^2 + 8x + 7 = (x + 1)(x + 7)$$
$$x^2 + 8x - 9 = (x - 1)(x + 9)$$

Tip:
Always look for a common factor first.

When guesswork becomes time-consuming, it may be quicker to use the method shown in the following example.

To factorise $12x^2 + 7x - 10$:

- Compare with $ax^2 + bx + c$: $a = 12, b = 7, c = -10$

- Calculate $a \times c$: $ac = -120$

- Find factors of ac that add to give b:

Factors of -120	Sum
120×-1	119
60×-2	58
20×-6	14
15×-8	7

Tip:
Since ac is negative and b is positive, the signs must be different and the 'larger' number must be positive.

Tip:
Keep practising and you will find that your guesses for the two numbers will become more efficient.

The factors of -120 that add to give 7 are 15 and -8.

- Write $7x$ as $15x - 8x$ (or as $-8x + 15x$):
$$12x^2 + 7x - 10 = 12x^2 + 15x - 8x - 10$$

- Group the terms in pairs, taking care with signs:
$$12x^2 + 7x - 10 = (12x^2 + 15x) - (8x + 10)$$

Tip:
Remember, if you put a bracket immediately after a minus, you must change the signs in the bracket.

- Take out any common factors from each of the pairs:
$$12x^2 + 7x - 10 = 3x(4x + 5) - 2(4x + 5)$$

- Finally take out a common expression:
$$12x^2 + 7x - 10 = (4x + 5)(3x - 2)$$

This method also works when the coefficient of x^2 is negative. However, to avoid problems with signs, it may be better to take out a factor of -1 first, as in Example 2.4.

Example 2.4 Factorise $f(x) = -9x^2 + 18x - 5$.

Step 1: Write $f(x)$ in the form $-g(x)$.
$$-9x^2 + 18x - 5 = -(9x^2 - 18x + 5)$$
Now consider $9x^2 - 18x + 5$.

Step 2: Identify a, b and c.
$a = 9, b = -18, c = 5$

Step 3: Find factors of ac that add up to give b.
$ac = 9 \times 5 = 45$

Factors of 45 that add to give -18 are -15 and -3.

Step 4: Rewrite bx using the factors found in Step 2.

Step 5: Factorise the first two and last two terms. Then take out the common bracket.

Step 6: Multiply *one* of the brackets by -1.

$$-9x^2 + 18x - 5 = -[9x^2 - 18x + 5]$$
$$= -[9x^2 - 15x - 3x + 5]$$
$$= -[(9x^2 - 15x) - (3x - 5)]$$
$$= -[3x(3x - 5) - 1(3x - 5)]$$
$$= -(3x - 5)(3x - 1)$$
$$= (3x - 5)(1 - 3x)$$

> **Tip:**
> Take out a factor of -1.

> **Tip:**
> It does not matter which bracket you multiply by -1, but do not multiply both. In this example $-(3x - 1) = -3x + 1$
> $= 1 - 3x$.

Important note:
In the special case when $a = 1$, the quadratic expression can be factorised very easily. In this case, ac is the same as c, so you just find two numbers that multiply to give c, the constant term, and add to give b, the x-term. These numbers then go in the brackets.

Consider $x^2 - x - 6$:

Factors of -6 that add to give -1 are -3 and 2.

So $x^2 - x - 6 = (x - 3)(x + 2)$.

2.5 Completing the square

Completing the square for a quadratic polynomial.

Remember the pattern when squaring linear functions:
$$(x + p)^2 = x^2 + 2px + p^2 \qquad (x - p)^2 = x^2 - 2px + p^2$$

Now rearrange these:
$$x^2 + 2px = (x + p)^2 - p^2 \qquad x^2 - 2px = (x - p)^2 - p^2$$

For example:
$$x^2 + 10x = (x + 5)^2 - 25 \qquad (p = 5)$$
$$x^2 - 6x = (x - 3)^2 - 9 \qquad (p = 3)$$
$$x^2 + 3x = (x + \tfrac{3}{2})^2 - \tfrac{9}{4} \qquad (p = \tfrac{3}{2})$$

> **Recall:**
> $(x + 5)^2 = x^2 + 10x + 25$
> $(x - 3)^2 = x^2 - 6x + 9$

> **Tip:**
> The number in the bracket is half the coefficient of x. Then subtract the square of this number.

This process is called **completing the square**. It can also be applied to quadratic expressions with a constant term, for example:
$$x^2 + 10x + 30 = (x + 5)^2 - 25 + 30 = (x + 5)^2 + 5$$
$$x^2 + 10x - 20 = (x + 5)^2 - 25 - 20 = (x + 5)^2 - 45$$

> **Note:**
> Use this technique to find the coordinates of the centre of a circle. See Section 3.4.

When $a \neq 1$:

$$3x^2 + 2x + 1 = 3(x^2 + \tfrac{2}{3}x) + 1$$
$$= 3[(x + \tfrac{1}{3})^2 - \tfrac{1}{9}] + 1$$
$$= 3(x + \tfrac{1}{3})^2 - \tfrac{1}{3} + 1$$
$$= 3(x + \tfrac{1}{3})^2 + \tfrac{2}{3}$$

Tip:

Take out a factor of 3 from the x^2 and x terms first.

When $a < 0$:

$$1 - 10x - x^2 = -[x^2 + 10x - 1]$$
$$= -[(x + 5)^2 - 25 - 1]$$
$$= -[(x + 5)^2 - 26]$$
$$= 26 - (x + 5)^2$$

Tip:

Take out a factor of -1 first. Remember to multiply through by it at the end.

Completing the square involves writing the quadratic expression $ax^2 + bx + c$ in the form $A(x + B)^2 + C$. You may prefer to complete the square using this identity, as in Example 2.5.

Example 2.5 Write $3x^2 + 2x + 1$ in the form $A(x + B)^2 + C$.

Step 1: Expand $A(x + B)^2 + C$.

$$A(x + B)^2 + C \equiv A(x + B)(x + B) + C$$
$$\equiv A(x^2 + 2Bx + B^2) + C$$
$$\equiv Ax^2 + 2ABx + AB^2 + C$$

Note:

The $\equiv$ symbol indicates that the expressions are identically equal, for all values of x.

Step 2: Compare coefficients with the original expression.

Step 3: Evaluate A, B and C.

So $3x^2 + 2x + 1 \equiv Ax^2 + 2ABx + AB^2 + C$

Coefficient of x^2: $3 = A$

Coefficient of x: $2 = 2AB \Rightarrow B = \tfrac{1}{3}$

Constant term: $1 = AB^2 + C \Rightarrow C = 1 - AB^2 = 1 - 3(\tfrac{1}{3})^2 = \tfrac{2}{3}$

Substituting $A = 3$, $B = \tfrac{1}{3}$ and $C = \tfrac{2}{3}$ gives

$$3x^2 + 2x + 1 \equiv 3(x + \tfrac{1}{3})^2 + \tfrac{2}{3}$$

Applications of completing the square

When a quadratic function $f(x)$ is in completed square form it is easy to find its maximum or minimum value and also the turning point and axis of symmetry of the curve $y = f(x)$.

In general, $f(x) = A(x + B)^2 + C$ has a turning point at $(-B, C)$. This is a minimum if $A > 0$ and a maximum if $A < 0$, for example:

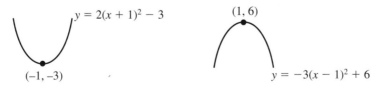

Tip:

If you write $y = -3(x - 1)^2 + 6$ as $y = 6 - 3(x - 1)^2$ it is easier to see that the maximum value is 6.

Example 2.6 It is given that $f(x) = 2x^2 - 4x + 5$.

  a Write $f(x)$ in the form $A(x + B)^2 + C$.

  b Write down the least value of $f(x)$ and state the value of x at which this occurs.

  c Hence state the coordinates of the minimum turning point on the curve $y = f(x)$.

  d Write down the equation of the axis of symmetry of the curve.

Step 1: Expand $A(x + B)^2 + C$.

a $2x^2 - 4x + 5 \equiv A(x + B)^2 + C$

$\equiv Ax^2 + 2ABx + AB^2 + C$

Step 2: Compare coefficients with the original expression, and evaluate A, B and C.

Coefficient of x^2: $\quad 2 = A$

Coefficient of x: $\quad -4 = 2AB \Rightarrow B = -1$

Constant term: $\quad 5 = AB^2 + C$

$\Rightarrow C = 5 - AB^2 = 5 - 2(-1)^2 = 3$

So $f(x) = 2x^2 - 4x + 5 = 2(x - 1)^2 + 3$

Step 3: Find the value of x such that $A(x + B)^2 = 0$.

Step 4: Substitute into $f(x)$.

b When $x = 1$, $2(x - 1)^2 = 0$,

so $f(x) = 0 + 3 = 3$.

For *all other values* of x,

$2(x - 1)^2 > 0 \Rightarrow f(x) > 3$.

Since $f(x) \geqslant 3$, the least value of $f(x)$ is 3 and it occurs when $x = 1$.

Tip:
Focus on the squared part of the expression.

Step 5: State $(-B, C)$.

c The curve $y = 2x^2 - 4x + 5$ has a minimum turning point at $(1, 3)$.

Note:
See Section 2.2 for the sketch of this curve.

Step 6: State the equation of the axis of symmetry.

d The axis of symmetry goes through the turning point so its equation is $x = 1$.

Tip:
See Translations (Section 3.8).

SKILLS CHECK 2B: Quadratic polynomials – factorising and completing the square

1 Factorise these expressions.

a $x^2 + 5x$ 　　　 **b** $x^2 - 2x + 1$ 　　　 **c** $a^2 - 16$

d $x^2 - 5x - 6$ 　　 **e** $x^2 + 13x - 30$ 　　 **f** $2x^2 - 8x$

2 Factorise these expressions.

a $2x^2 + 7x + 6$ 　　 **b** $5x^2 - 14x - 3$ 　　 **c** $x^3 - 4x^2 - 21x$

d $3y^2 + 4y - 4$ 　　 **e** $12 - 4x - 40x^2$ 　　 **f** $4x^2 - 25$

3 Write each of these quadratic polynomials in the form $A(x + B)^2 + C$.

a $x^2 + 6x + 8$ 　　 **b** $x^2 - 12x - 3$ 　　 **c** $x^2 + 5x - 2$

4 Write the quadratic polynomials in question **2** in the form $A(x + B)^2 + C$.

5 It is given that $x^2 - 4x + 7 = (x - p)^2 + q$.

　a Find p and q.

　b Hence write down the coordinates of the vertex of the curve $y = x^2 - 4x + 8$.

　c State whether the vertex is a maximum or a minimum turning point.

　d Write down the equation of the axis of symmetry of the curve.

6 It is given that $f(x) = 4x^2 + 8x + 1$.

　a Write $f(x)$ in the form $A(x + B)^2 + C$.

　b Find the coordinates of the vertex of the curve $y = f(x)$, stating whether it is a maximum or a minimum turning point.

　c Write down the equation of the axis of symmetry of the curve.

7 **a** Write $f(x) = 14 - 4x - x^2$ in the form $C - (x + B)^2$.

　b Hence state the maximum value of $f(x)$.

　c Write down the coordinates of the vertex of the curve $y = 14 - 4x - x^2$.

SKILLS CHECK 2B EXTRA is on the CD

Solve quadratic equations.

A quadratic equation has the form $ax^2 + bx + c = 0$, where $a \neq 0$. It can have two real solutions or one real solution or no real solutions.

There are several ways of solving quadratic equations. Here are three important techniques.

Solving by factorising

If the quadratic expression can be factorised into the product of two linear functions, solve the quadratic equation using the fact that if $p \times q = 0$, then either $p = 0$ or $q = 0$.

Note:
Not all quadratic expressions can be factorised.

Example 2.7   **a**   Factorise $f(x) = 6x^2 + 11x + 3$.

b   Hence solve $6x^2 + 11x + 3 = 0$.

Step 1: Factorise the quadratic expression.

a $f(x) = 6x^2 + 11x + 3$

$\qquad = 6x^2 + 9x + 2x + 3$

$\qquad = (6x^2 + 9x) + (2x + 3)$

$\qquad = 3x(2x + 3) + 1(2x + 3)$

$\qquad = (2x + 3)(3x + 1)$

Working for factorisation:
$a = 6, b = 11, c = 3$
$ac = 18$
Factors of 18 that add to give 11 are 9 and 2.

Recall:
Section 2.4 for method. Alternatively, factorise by guesswork.

b $\qquad 6x^2 + 11x + 3 = 0$

Step 2: Use $pq = 0 \Rightarrow$ $p = 0$ or $q = 0$ and solve the resulting linear equations.

$\Rightarrow (2x + 3)(3x + 1) = 0$

$\Rightarrow 2x + 3 = 0 \qquad$ or $\qquad 3x + 1 = 0$

$\qquad x = -\frac{3}{2} \qquad\qquad\qquad x = -\frac{1}{3}$

Note:
It is usual to leave answers as 'top heavy' fractions if necessary.

Example 2.8   Solve $8x^2 + 4x - 2 = -x(2x - 3)$.

Step 1: Rearrange the equation to $f(x) = 0$.

$8x^2 + 4x - 2 = -2x^2 + 3x$

$10x^2 + x - 2 = 0$

Step 2: Factorise the quadratic expression.

$(2x + 1)(5x - 2) = 0$

$\Rightarrow 2x + 1 = 0 \qquad$ or $\qquad 5x - 2 = 0$

Step 3: Use $pq = 0 \Rightarrow$ $p = 0$ or $q = 0$ and solve the resulting linear equations.

$\qquad x = -\frac{1}{2} \qquad\qquad\qquad x = \frac{2}{5}$

Tip:
This expression is easy to factorise by guesswork.

Solving by completing the square

Complete the square for the quadratic expression and then rearrange to make x the subject.

Example 2.9   **a**   Write $x^2 - 8x + 9$ in the form $(x - p)^2 - q$.

b   Hence solve the equation $x^2 - 8x + 9 = 0$, leaving your answers in surd form.

Step 1: Complete the square

a $x^2 - 8x + 9 = (x - 4)^2 - 16 + 9$

$\qquad\qquad\qquad = (x - 4)^2 - 7$

Recall:
Completing the square (Section 2.5).

b
$$x^2 - 8x + 9 = 0$$
$$\Rightarrow (x - 4)^2 - 7 = 0$$

Step 2: Make x the subject.　　(Add 7)　　　　　$(x - 4)^2 = 7$

(Square root both sides)　$x - 4 = \pm\sqrt{7}$

(Add 4)　　　　　　　$x = 4 \pm\sqrt{7}$

The two solutions are $x = 4 + \sqrt{7}$ and $x = 4 - \sqrt{7}$.

Example 2.10　**a**　Write $4x^2 + 4x - 15$ in the form $A(x + B)^2 + C$.

　　b　Hence solve $4x^2 + 4x - 15 = 0$.

Step 1: Complete the square.　**a**　Let $4x^2 + 4x - 15 \equiv A(x + B)^2 + C$
$$\equiv Ax^2 + 2ABx + AB^2 + C$$

Coefficient of x^2:　　　$4 = A$

Coefficient of x:　　　$4 = 2AB \Rightarrow B = \frac{1}{2}$

Constant term:　　　$-15 = AB^2 + C \Rightarrow C = -15 - 4(\frac{1}{2})^2 = -16$

$4x^2 + 4x - 15 = 4(x + \frac{1}{2})^2 - 16$

Step 2: Make x the subject.　**b**　Rearrange:　　$4(x + \frac{1}{2})^2 = 16$

Divide by 4:　　$(x + \frac{1}{2})^2 = 4$

Take the square root of each side:

$$x + \frac{1}{2} = \pm 2$$

$$x = -\frac{1}{2} \pm 2$$

Either　$x = -\frac{1}{2} - 2$　or　$x = -\frac{1}{2} + 2$

　　　$= -\frac{5}{2}$　　　　　$= \frac{3}{2}$

Using the quadratic formula

If $ax^2 + bx + c = 0$ then

$$x = \frac{-b \pm \sqrt{b^2 - 4ac}}{2a}$$

You must memorise this formula for your examination.

Example 2.11　Use the quadratic formula to solve $2x^2 + x - 4 = 0$, leaving your answers in surd form.

Step 1: Identify a, b and c.　$a = 2$, $b = 1$, $c = -4$

Step 2: Substitute into the quadratic formula and evaluate.
$$x = \frac{-b \pm \sqrt{b^2 - 4ac}}{2a}$$

$$= \frac{-1 \pm \sqrt{1^2 - 4(2)(-4)}}{2 \times 2}$$

$$= \frac{-1 \pm \sqrt{33}}{4}$$

$$x = \frac{-1 + \sqrt{33}}{4} \quad \text{or} \quad \frac{-1 - \sqrt{33}}{4}$$

Tip:
Bring x to one side and all other terms to the other side.

Tip:
Remember to include $\pm$ when taking the square root. This gives two distinct roots.

Tip:
Bring x to one side and all other terms to the other side.

Tip:
Remember to include $\pm$ when taking the square root.

Note:
The formula is derived by completing the square on $ax^2 + bx + c = 0$.

Note:
These are exact answers. In this case, the values given by a calculator would be approximations.

Number of solutions of a quadratic equation

The term under the square root in the quadratic formula is the **discriminant**, $b^2 - 4ac$. Its value can be used to deduce the number of solutions of the quadratic equation $ax^2 + bx + c = 0$.

Recall:
The discriminant (Section 2.3).

If $b^2 - 4ac > 0$, then $\sqrt{b^2 - 4ac}$ can be calculated and the quadratic formula will give **two different values** for x. There are two solutions. We say that the quadratic equation has **two real distinct roots**.

Tip:
Learn these conditions.

If $b^2 - 4ac = 0$, then $\sqrt{b^2 - 4ac} = 0$ and the quadratic formula will give **only one value** for x. There is one solution.
The quadratic equation has two equal roots (**repeated roots**).

If $b^2 - 4ac < 0$, then there are no real values of $\sqrt{b^2 - 4ac}$. There are no solutions and we say that the quadratic equation has **no real roots**.

Note:
See Section 2.8 for more on these conditions.

Example 2.12 The equation $2x^2 - 3x + 3k = 0$ has equal roots. Find the value of k.

Step 1: Identify a, b and c.　$a = 2, b = -3, c = 3k$

Step 2: Find $b^2 - 4ac$ and use the appropriate condition for the discriminant.

$$b^2 - 4ac = (-3)^2 - 4(2)(3k)$$
$$= 9 - 24k$$

Equal roots $\Rightarrow b^2 - 4ac = 0$
$$9 - 24k = 0$$
$$k = \frac{9}{24}$$
$$= \frac{3}{8}$$

SKILLS CHECK 2C: Quadratic equations

1   Solve the following quadratic equations:

   a $(x - 2)(x + 3) = 0$ **b** $(1 - 4x)(3 + 2x) = 0$ **c** $4x(x + 5) = 0$

2   Solve the following equations, using the method of factorisation.

   a $x^2 + 6x + 5 = 0$ **b** $x^2 - 11x + 24 = 0$ **c** $x^2 - 6x = 0$

   d $x^2 - 5x - 6 = 0$ **e** $x^2 - x - 6 = 0$ **f** $x^2 - 36 = 0$

3   The function f is defined for all x by $f(x) = x^2 + 3x - 5$.

   a   Express $f(x)$ in the form $(x + P)^2 + Q$.

   b   Hence, or otherwise, solve the equation $f(x) = 0$, giving your answers in surd form.

4   **a**   Write $2x^2 - 3x - 2$ in the form $A(x + B)^2 + C$.

   b   Hence solve $2x^2 - 3x - 2 = 0$.

   c   Check your answers by solving $2x^2 - 3x - 2 = 0$ by the method of factorisation.

5   Use the quadratic formula to solve the equation $5x^2 + 3x - 3 = 0$, leaving your answers in surd form.

6   **a**   Solve $(2x - 3)^2 = 25$.

   b   Solve $(2x - 3)^2 = 2x$, expressing your answers in surd form.

7 Find the exact solutions of the quadratic equation $3x^2 + 2x - 4 = 0$.

8 By calculating the discriminant, find the number of real solutions of each of the following quadratic equations:

    a $x^2 - 3x + 1 = 0$ **b** $2x^2 - 3x - 1 = 0$ **c** $4x^2 - 4x + 1 = 0$ **d** $5x + x^2 - 3 = 0$

9 Find the discriminant of $3x^2 - 2x + 5$ and hence show that $3x^2 - 2x + 5 = 0$ has no real solutions.

SKILLS CHECK **2C EXTRA** is on the CD

2.7 Simultaneous equations

Solve by substitution a pair of simultaneous equations of which one is linear and the other is quadratic.

Linear simultaneous equations

In a pair of simultaneous equations there are two unknowns, for example x and y.

Consider these simultaneous equations.

$2y + 3x = 18$ ①
$5y - x = 11$ ②

To **solve** them you have to find a value of x and a value of y that satisfy *both* equations. The most common methods to use are elimination and substitution.

Note:
You may have to solve linear simultaneous equations in work on coordinate geometry in Chapter 3.

Elimination method

When the numerical coefficients of one of the unknowns are the same, *eliminate* that unknown either by adding or subtracting the equations.

Step 1: Make the coefficients of one of the unknowns the same then add or subtract to eliminate the unknown.

$② \times 3$ $15y - 3x = 33$ ③
            $2y + 3x = 18$ ①
$③ + ①$ $17y \qquad = 51$
                $y = 3$

Step 2: Substitute the value found into one of the original equations.

Substituting $y = 3$ into equation ①:

$6 + 3x = 18$
    $3x = 12 \quad \Rightarrow \quad x = 4$

The solution is $x = 4$, $y = 3$.

Note:
If the terms containing the unknown you want to eliminate have the same sign, subtract. If the signs are different, add.

Note:
The lines $2y + 3x = 18$ and $5y - x = 11$ intersect at $(4, 3)$. See Section 3.6 for more on graphical interpretations.

Substitution method

Use one equation to express one of the unknowns in terms of the other and then *substitute* for it in the other equation.

Step 1: Express one unknown in terms of the other.

Write x in terms of y using equation ②:

$x = 5y - 11$

Step 2: Substitute it into the other equation and solve.

Substituting for x in equation ①:

$2y + 3(5y - 11) = 18$
   $2y + 15y - 33 = 18$
         $17y = 51 \quad \Rightarrow \quad y = 3$

Step 3: Substitute the value found into one of the original equations.

To find x, proceed as in the elimination method.

Tip:
Although it does not matter whether you write x in terms of y, or y in terms of x, try to avoid expressions with fractions where possible.

Quadratic inequalities

Inequalities such as $x^2 < 9$ and $x^2 \geq 25$ are the simplest quadratic inequalities to solve.

a To solve $x^2 < 9$, first consider $x^2 = 9$.

$x^2 = 9$ when $x = 3$ or $x = -3$.

For any value between -3 and 3, $x^2 < 9$.

So the range of values of x for which $x^2 < 9$ is $-3 < x < 3$.

b To solve $x^2 \geq 25$, first consider $x^2 = 25$.

$x^2 = 25$ when $x = 5$ or $x = -5$.

If $x \leq -5$, $x^2 \geq 25$. Also if $x \geq 5$, $x^2 \geq 25$.

So the values of x for which $x^2 \geq 25$ are $x \leq -5$ or $x \geq 5$.

This method can be applied to more complicated quadratic inequalities by completing the square.

Tip:
In **a** the values are 'sandwiched' between -3 and 3, so the solution should be written in one inequality.

Tip:
In **b** the values are outside the 'sandwich', so give two separate inequalities.

Recall:
Completing the square (Section 2.5).

Example 2.16

a Express $x^2 - 6x + 7$ in the form $(x + a)^2 + b$.

b Find the range of values of x for which $x^2 - 6x + 7 < 0$.

Step 1: Complete the square.

a $x^2 - 6x + 7 = (x - 3)^2 - 9 + 7$
$\qquad\qquad\qquad = (x - 3)^2 - 2$

b $\quad x^2 - 6x + 7 < 0$

Step 2: Solve the inequality using the completed square format.

$\Rightarrow (x - 3)^2 - 2 < 0$
$\qquad (x - 3)^2 < 2$
$\qquad -\sqrt{2} < x - 3 < \sqrt{2}$
$(+3) \quad 3 - \sqrt{2} < x < 3 + \sqrt{2}$

So $x^2 - 6x + 7 < 0$ when $3 - \sqrt{2} < x < 3 + \sqrt{2}$.

Tip:
Substitute y for $(x - 3)$ and use the fact that if $y^2 < 2$, then $-\sqrt{2} < y < \sqrt{2}$.

Recall:
Surds (Section 1.2).

Another method is to use a sketch. This is illustrated in Example 2.17.

Example 2.17

It is given that $f(x) = x^2 + 2x - 8$.

a Solve $f(x) = 0$.

b Sketch $y = f(x)$.

c Hence solve $x^2 + 2x - 8 \geq 0$.

Step 1: Solve the quadratic equation.

a $f(x) = 0$ when $x^2 + 2x - 8 = 0$
$\qquad\qquad\qquad (x - 2)(x + 4) = 0$
$\Rightarrow \quad x = 2, -4$.

Step 2: State where the curve crosses the x-axis.

The curve $y = f(x)$ crosses the x-axis at $(2, 0)$ and $(-4, 0)$.

Step 3: Sketch the curve.

b

Note:
Sketching a quadratic curve is described more fully in Section 3.7.

Recall:
Solution of quadratic equations (Section 2.6).

Recall:
The coefficient of x^2 is positive, so the parabola is $\cup$-shaped (Section 2.2).

Tip:
You can use 'filled in' circles to show that x can be 2 or -4.

Step 4: Indicate the possible x-values on the sketch and solve $f(x) \geq 0$.

c $f(x) \geq 0$ when the curve is on or above the x-axis, so from the sketch, $f(x) \geq 0$ when $x \leq -4, x \geq 2$.

Tip:
Write this solution in two separate inequalities.

Application of inequalities to roots of equations

You may need to solve a quadratic inequality in questions about roots of equations, as in the following example.

Example 2.18 Find the values of k for which $2x^2 - kx + 2 = 0$ has no real roots.

Step 1: Identify a, b and c.

Comparing $2x^2 - kx + 2$ with $ax^2 + bx + c$,

$a = 2$, $b = -k$, $c = 2$

If the equation has no real roots, then

$$b^2 - 4ac < 0$$

Step 2: Find $b^2 - 4ac$, and use the appropriate condition for the discriminant.

$\Rightarrow \quad (-k)^2 - 4 \times 2 \times 2 < 0$

$$k^2 - 16 < 0$$

Let $f(k) = k^2 - 16$

$= (k - 4)(k + 4)$

$f(k) = 0$ when $k = 4$ or -4

Step 3: Solve the inequality in k by sketching $y = f(k)$ and finding the values of k when the curve is below the k-axis.

The graph of $y = f(k)$ goes through $(-4, 0)$ and $(4, 0)$.

From the sketch, $f(k) < 0$ when $-4 < k < 4$.

So $2x^2 - kx + 2 = 0$ has no real roots when $-4 < k < 4$.

> **Recall:**
> Conditions to be satisfied by the discriminant (Section 2.6).

> **Tip:**
> You can use 'open' circles to show that k cannot take the values 4 and -4.

> **Note:**
> This solution can be written in one statement.

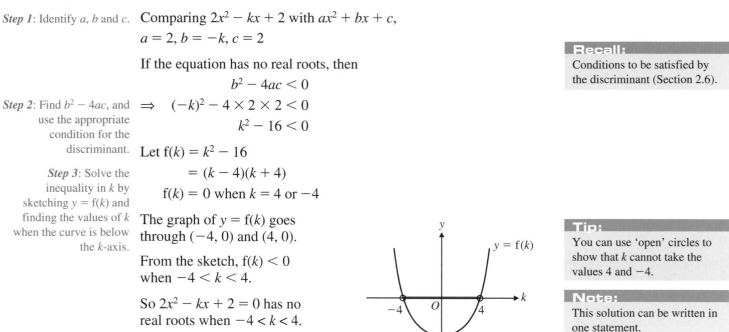

SKILLS CHECK **2E: Inequalities**

1 Solve these linear inequalities.

a $4x - 5 > 7$ **b** $3 - 5x < 8$ **c** $2(3x - 1) \geqslant 3(x + 8)$

d $5y - (4 + y) > 0$ **e** $4 < \dfrac{2x}{3}$ **f** $\dfrac{1}{2}x + \dfrac{3}{4} \leqslant 8$

2 Solve these inequalities:

a $5 < 2x - 1 < 17$ **b** $-3 \leqslant \dfrac{x}{2} \leqslant 5$

 3 The solution of the inequality $\sqrt{3}\,(x + \sqrt{3}) > 6$ is $x > \sqrt{a}$. Find the value of a.

4 Solve these quadratic inequalities.

a $y^2 > 4$ **b** $x^2 \leqslant 49$ **c** $x^2 \geqslant 5$

d $2x^2 < 18$ **e** $(x - 1)^2 > 4$ **f** $(x + 2)^2 \leqslant 5$

5 a Express $x^2 + 4x - 5$ in the form $(x + p)^2 + q$, finding the values of the constants p and q.

b Find the values of x for which $x^2 + 4x \geqslant 5$.

6 Solve these quadratic inequalities.

  a $(x + 4)(x - 3) < 0$ **b** $3(2x + 5)(3x - 2) \geqslant 0$

  c $(4 - x)(5 + x) \leqslant 0$ **d** $p^2 + 7p + 10 < 0$

  e $2x^2 + x \geqslant 6$

7 a The quadratic equation $x^2 - 4x - 6 = 0$ has solutions $x = p \pm \sqrt{q}$ where p and q are integers. Find the values of p and q.

  b Sketch $y = x^2 - 4x - 6$.

  c Solve the inequality $x^2 - 4x - 6 < 0$.

8 a Find the values of k for which the equation $3x^2 - kx + 3 = 0$ has no real roots.

  b The equation $kx^2 - 8x + k = 0$ has real roots. Show that $-4 \leqslant k \leqslant 4$.

9 a Find the discriminant of the quadratic expression $2x^2 - kx + 2$.

  b Find the values of k for which $2x^2 - kx + 2 = 0$ has two distinct real roots.

SKILLS CHECK **2E EXTRA** is on the CD

2.9 Disguised quadratic equations

Recognise and solve equations in x which are quadratic in some function of x,
e.g. $x^{\frac{2}{3}} - 5x^{\frac{1}{3}} + 4 = 0$.

Sometimes equations occur that can be solved as quadratic equations even though they are not always of the form $ax^2 + bx + c = 0$.

Example 2.19 Solve the equation $x^4 - 10x^2 + 9 = 0$.

Step 1: Decide on the function of x that will make the equation into a quadratic.

Let $y = x^2$.

The equation is now $\quad y^2 - 10y + 9 = 0$

> **Tip:**
> Use another letter for the function to avoid confusion.

Step 2: Rewrite the equation and solve by the appropriate method.

Factorise $\quad (y - 9)(y - 1) = 0$

Solve each bracket $\quad y = 9$ or $y = 1$

Step 3: Substitute the original function of x and solve.

Substitute and solve for x $\quad x^2 = 9 \Rightarrow x = \pm 3$

$\qquad\qquad\qquad\qquad\quad x^2 = 1 \Rightarrow x = \pm 1$

$\qquad\qquad\qquad\qquad\quad x = \pm 3$ or $x = \pm 1$

Example 2.20 Solve the equation $x^{\frac{2}{3}} - 5x^{\frac{1}{3}} + 4 = 0$.

Step 1: Decide on the function of x that will make the equation into a quadratic.

Let $y = x^{\frac{1}{3}}$.

The equation is now $\quad y^2 - 5y + 4 = 0$

Step 2: Rewrite the equation and solve by the appropriate method.

Factorise $\quad (y - 4)(y - 1) = 0$

Solve each bracket $\quad y = 4$ or $y = 1$

Step 3: Substitute the original function of x and solve.

Substitute and solve for x $\quad x^{\frac{1}{3}} = 4 \Rightarrow x = 64$

$\qquad\qquad\qquad\qquad\quad x^{\frac{1}{3}} = 1 \Rightarrow x = 1$

$\qquad\qquad\qquad\qquad\quad x = 64$ or $x = 1$

> **Tip:**
> Cube each side.

1 Solve the following equations by using the given substitution.

 a $x + x^{\frac{1}{2}} - 6 = 0$ using $y = x^{\frac{1}{2}}$

 b $x^6 - 9x^3 + 8 = 0$ using $y = x^3$

2 Solve the following equations using an appropriate substitution.

 a $x^4 - 13x^2 + 36 = 0$

 b $x^{\frac{2}{3}} - 4x^{\frac{1}{3}} - 5 = 0$

 c $3x + 2x^{\frac{1}{2}} - 8 = 0$

 d $9x^{\frac{4}{3}} - 13x^{\frac{2}{3}} + 4 = 0$

SKILLS CHECK **2F EXTRA is on the CD**

Examination practice Polynomials

1 i Find the constants a, b and c such that for all values of x,

$$4x^2 + 40x + 97 = a(x + b)^2 + c.$$

 ii Hence write down the equation of the line of symmetry of the
 curve $y = 4x^2 + 40x + 97$. [OCR May 2003]

2 The quadratic equation $x^2 + kx + k = 0$ has no real roots for x.

 i Write down the discriminant of $x^2 + kx + k$ in terms of k.

 ii Hence find the set of values that k can take. [OCR Spec paper]

3 i Calculate the discriminant of the quadratic polynomial $-x^2 + 4x - 7$.

 ii State the number of real roots of the equation $-x^2 + 4x - 7 = 0$ and
 hence explain why $-x^2 + 4x - 7 < 0$ for all values of x. [OCR Jan 2004]

4 By susbstituting $t = x^{\frac{1}{2}}$, find the values of x for which $2x + 3 = 7x^{\frac{1}{2}}$. [OCR May 2004]

5 i Find the constants a and b such that, for all values of x,

$$x^2 + 6x + 20 = (x + a)^2 + b.$$

 ii Hence state the least value of $x^2 + 6x + 20$, and state also the value of x
 for which this least value occurs.

 iii Write down the greatest value of $\dfrac{1}{x^2 + 6x + 20}$. [OCR Jan 2003]

6 i Express $3x^2 + 4x + 1$ in the form $a(x + b)^2 + c$.

  ii Hence, or otherwise, find the coordinates of the vertex of the graph of $y = 3x^2 + 4x + 1$.   [OCR May 2004]

7 i Given that $\sqrt{x} = y$, show that the equation

$$\sqrt{x} + \frac{10}{\sqrt{x}} = 7$$

  may be written as

$$y^2 - 7y + 10 = 0.$$

  ii Hence solve the equation

$$\sqrt{x} + \frac{10}{\sqrt{x}} = 7$$ [OCR Jan 2003]

8 Solve the inequality $4x - 3x^2 > 0$.   [OCR May 2004]

9 Solve the inequality $2 + 3(x - 4) < 3(2x - 5)$.   [OCR Jan 2004]

10 Solve the simultaneous equations

$$x^2 + 5x + y = 4, \quad x + y = 8.$$ [OCR May 2004]

11 Expand $(x + 4)(x - 5)(2x - 3)$, simplifying your answer.   [OCR June 2004]

12 Solve the simultaneous equations

$$x + y = 2, \quad x^2 + 2y^2 = 11.$$ [OCR Nov 2003]

13 By substituting $y = x^2$, or otherwise, find the real roots of the equation $x^4 - 3x^2 - 4 = 0$.   [OCR Nov 2003]

14 Express $2x^2 + 12x + 13$ in the form $a(x + b)^2 + c$.   [OCR Nov 2003]

3 Coordinate geometry and graphs

3.1 Line segments

Find the length, gradient and midpoint of a line segment, given the coordinates of its end points.

Midpoint

The midpoint of the line segment joining $A(x_1, y_1)$ and $B(x_2, y_2)$ is given by the formula

$$\text{Midpoint} = \left(\frac{x_1 + x_2}{2}, \frac{y_1 + y_2}{2}\right)$$

Tip:
The order in which the coordinates are added does not matter, but make sure you add the two x-coordinates together and the two y-coordinates together.

Example 3.1 Find the midpoint of AB where A is the point $(7, -3)$ and B is the point $(-2, -5)$.

Step 1: Substitute in the midpoint formula.

Taking $(7, -3)$ as (x_1, y_1) and $(-2, -5)$ as (x_2, y_2):

$$\text{Midpoint} = \left(\frac{x_1 + x_2}{2}, \frac{y_1 + y_2}{2}\right)$$

$$= \left(\frac{7 + (-2)}{2}, \frac{(-3) + (-5)}{2}\right)$$

$$= \left(\frac{5}{2}, -4\right)$$

Tip:
Drawing a sketch can help you to see whether your answer is reasonable.

Tip:
Be very careful with negatives.

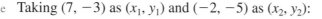

Distance between two points

The distance AB between two points $A(x_1, y_1)$ and $B(x_2, y_2)$ is given by the formula

$$AB = \sqrt{(x_2 - x_1)^2 + (y_2 - y_1)^2}$$

Tip:
Work out the brackets before trying to square them. Remember that squares can never be negative.

Example 3.2 Find the distance between A and B where A is the point $(7, -3)$ and B is the point $(-2, -5)$.

Step 1: Substitute into the formula.
Step 2: Work out each bracket, square, add, then square root.

Taking $(7, -3)$ as (x_1, y_1) and $(-2, -5)$ as (x_2, y_2):

$$AB = \sqrt{(-2 - 7)^2 + (-5 - (-3))^2}$$

$$= \sqrt{(-9)^2 + (-2)^2}$$

$$= \sqrt{85}$$

Note:
You will not be allowed a calculator in Core 1, so if the answer is not a perfect square, leave it in surd form.

Example 3.3 P is the point $(-6, 1)$ and Q is the point $(8, -1)$. The length PQ is $k\sqrt{2}$. Find the value of k.

Step 1: Substitute into the formula.
Step 2: Work out each bracket, square, add, then square root.

Taking $(-6, 1)$ as (x_1, y_1) and $(8, -1)$ as (x_2, y_2):

$$AB = \sqrt{(8 - (-6))^2 + (-1 - 1)^2}$$

$$= \sqrt{14^2 + (-2)^2}$$

$$= \sqrt{200}$$

Step 3: Simplify the surd.

$$= \sqrt{100 \times 2}$$

$$= 10\sqrt{2}$$

Therefore $k = 10$.

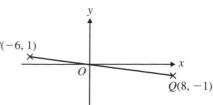

25

Gradient

The gradient of the line joining $A(x_1, y_1)$ and $B(x_2, y_2)$ is given by the formula

$$\text{Gradient of } AB = \frac{y_2 - y_1}{x_2 - x_1}$$

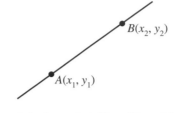

Tip:
Keep the order the same and be careful with minuses.

Example 3.4 Find the gradient of the line AB where A is the point $(7, -3)$ and B is the point $(-2, -5)$.

Step 1: Substitute into the formula.

$$\text{Gradient of } AB = \frac{-5 - (-3)}{-2 - 7} = \frac{-2}{-9} = \frac{2}{9}$$

Tip:
Substitute into the formula before attempting to work anything out.

3.2 Equations of lines

Find the equation of a straight line given sufficient information (e.g. the coordinates of two points on it or one point on it and its gradient). Interpret and use linear equations, particularly the forms $y = mx + c$, $y - y_1 = m(x - x_1)$ and $ax + by + c = 0$.

There are several ways of writing the equation of a straight line.

Gradient–intercept format
The equation of the line is $y = mx + c$,
where m is the gradient
and c is the y-intercept.

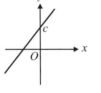

Note:
The y-intercept is the y-coordinate of the point where the line crosses the y-axis.

For example, the line $y = 3x - 2$ has gradient 3 and crosses the y-axis at $(0, -2)$.

The format $ax + by + c = 0$, where a, b and c are integers

Example 3.5 Write the equation of the following straight line in the form $ax + by + c = 0$, where a, b and c are integers.

Step 1: Eliminate any fractions.

$$y = -\tfrac{1}{4}x - \tfrac{5}{6}$$

Step 2: Collect all terms on one side, with zero on the other.

$$(\times 12) \qquad 12y = -3x - 10$$

$$3x + 12y + 10 = 0$$

Tip:
To eliminate the fractions multiply *every* term by the lowest common denominator (12 in this case).

The format $y - y_1 = m(x - x_1)$
Take P to be the general point (x, y).
A is the fixed point (x_1, y_1).

Using the gradient formula (calling the gradient m):

$$m = \frac{y - y_1}{x - x_1}$$

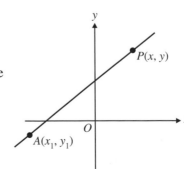

Note:
This form is useful when you know the gradient and a point on the line.

This can be rearranged to give

$$y - y_1 = m(x - x_1)$$

Example 3.6 Find the equation of the line with gradient $\frac{2}{3}$ that passes through the point $(-3, 1)$. Write the answer in the form $ax + by + c = 0$, where a, b and c are integers.

Step 1: Substitute into the formula.

$m = \frac{2}{3}$ and $(x_1, y_1) = (-3, 1)$.

The equation of the line is
$$y - y_1 = m(x - x_1)$$

Step 2: Multiply by the denominator if necessary.

$$y - 1 = \frac{2}{3}(x - (-3))$$
$$3(y - 1) = 2(x + 3)$$

Step 3: Rearrange into the required format.

$$3y - 3 = 2x + 6$$
$$-2x + 3y - 9 = 0$$

> **Note:**
> You could write
> $2x - 3y + 9 = 0$ to avoid starting with a minus sign.

Example 3.7 The points A and B have coordinates $(12, 5)$ and $(7, 3)$. Find:

a the gradient of AB

b the equation of the line AB.

Step 1: Use the gradient formula.

a Gradient $= \dfrac{3 - 5}{7 - 12} = \dfrac{-2}{-5} = \dfrac{2}{5}$

Step 2: Use the known gradient, known point formula.

b Equation of line AB is
$$y - y_1 = m(x - x_1)$$
$$y - 5 = \frac{2}{5}(x - 12)$$

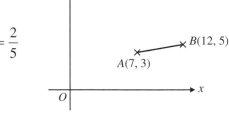

> **Tip:**
> It does not matter which point you use as (x_1, y_1).

> **Tip:**
> If you are not asked to give a specific format, then you may leave it like this.

You may have to find the **point of intersection** of two lines. In this case, solve the two equations simultaneously.

Example 3.8 Find the point of intersection of the two lines

$y = \frac{2}{3}x - 4$ and $4x - y = 9$.

$$y = \frac{2}{3}x - 4 \qquad ①$$
$$4x - y = 9 \qquad ②$$

> **Recall:**
> Simultaneous equations (Section 2.7) and points of intersection (Section 3.6).

Step 1: Substitute for one variable from one equation into the other.

Substitute for y from ① into ②:

$$4x - (\tfrac{2}{3}x - 4) = 9$$

Step 2: Simplify and solve the resulting equation.

$$4x - \tfrac{2}{3}x + 4 = 9$$
$$4x - \tfrac{2}{3}x = 5$$

Multiply through by 3:
$$12x - 2x = 15$$
$$x = \tfrac{3}{2}$$

Step 3: Substitute the value found into one of the original equations.

Substitute for x in ②:

$$6 - y = 9 \quad \Rightarrow \quad y = -3$$

The lines intersect at $(\tfrac{3}{2}, -3)$.

1 Write the following lines in the general form $ax + by + c = 0$.

a $y = 3x - 2$ **b** $x + \frac{2}{3}y = 2$ **c** $\frac{1}{2}x + \frac{3}{4}y = 5$ **d** $\frac{3}{5}x = 2 - \frac{1}{2}y$

2 a Write the straight line $4x - 5y = 8$ in the form $y = mx + c$.

  b What is the gradient and point of intercept on the y-axis of the straight line $2x + 3y = 6$?

3 For the following pairs of points, calculate **i** the gradient of AB, **ii** the midpoint of AB and **iii** the distance AB, writing your answer in surd form if necessary.

  a $A(-3, 6), B(4, -1)$ **b** $A(4, 6), B(-2, -4)$ **c** $A(-1, -2), B(2, -1)$

4 Find the equation of the line that has:

  a a gradient of $\frac{3}{5}$ passing through the point $(-4, -2)$

  b a gradient of -3 passing through the point $(5, -3)$

  c a line passing through the points $(-3, 6)$ and $(4, -1)$. (*Hint: work out the gradient first.*)

5 A triangle is formed by three straight lines, $y = \frac{1}{2}x$, $2x + y + 5 = 0$ and $x + 3y - 5 = 0$.

Prove that the triangle is isosceles.

6 In the triangle ABC, A, B and C are the points $(-4, 1)$, $(-2, -3)$ and $(3, 2)$, respectively.

  a Show that ABC is isosceles.

  b Find the coordinates of the midpoint of the base.

  c Find the area of ABC.

SKILLS CHECK **3A EXTRA** is on the **CD**

3.3 Parallel and perpendicular lines

Understand and use the relationships between the gradients of parallel and perpendicular lines.

For two lines with gradients m_1 and m_2:

the lines are **parallel** if their gradients are the same, i.e. $m_1 = m_2$;

the lines are **perpendicular** if the product of their gradients is -1, i.e. $m_1 \times m_2 = -1$.

Note:
Perpendicular lines have gradients that are negative reciprocals of each other, i.e.
$$m_1 = -\frac{1}{m_2}.$$

Example 3.9   Show that the lines $y = \frac{2}{3}x - 1$ and $2x - 3y + 6 = 0$ are parallel.

Step 1: Rearrange the equations of the lines into the form $y = mx + c$.

The first line has gradient $\frac{2}{3}$.

Rearranging the equation of the second line:
$$2x - 3y + 6 = 0$$
$$3y = 2x + 6$$
$$y = \frac{2}{3}x + 2$$

The second line has gradient $\frac{2}{3}$.

Recall:
When written in the form $y = mx + c$, the gradient is m.

Step 2: Compare gradients.   Since both lines have the same gradient, they are parallel.

Example 3.10 Find the equation of the line passing through $(0, 2)$ that is perpendicular to the line $y = \frac{1}{4}x + 3$.

Step 1: Find the gradient of the required line.

The gradient of the given line is $\frac{1}{4}$.

The gradient of the required line is the negative reciprocal of $\frac{1}{4}$, so the gradient of the required line is -4.

Step 2: Use the gradient and known point to find the equation of the line.

The given point $(0, 2)$ is the intercept on the y-axis, so the equation of the required line is $y = -4x + 2$.

> **Tip:**
> Unless a particular equation is asked for, you can give any form.

Example 3.11 Find the equation of the perpendicular bisector of the line AB where A is the point $(-3, 5)$ and B is the point $(-1, 3)$.

Step 1: Find the gradient of AB.

$$\text{Gradient of } AB = \frac{y_2 - y_1}{x_2 - x_1} = \frac{3 - 5}{-1 - (-3)} = \frac{-2}{2} = -1$$

Step 2: Find the gradient of a line perpendicular to AB.

Gradient of perpendicular bisector $= 1$

Step 3: Find the midpoint of AB.

Midpoint of AB

$$= \left(\frac{x_1 + x_2}{2}, \frac{y_1 + y_2}{2}\right)$$

$$= \left(\frac{(-3) + (-1)}{2}, \frac{5 + 3}{2}\right)$$

$$= (-2, 4)$$

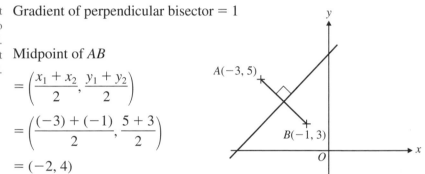

> **Recall:**
> Product of gradients is -1 for perpendicular lines.

> **Tip:**
> It may be helpful to draw a sketch.

Step 4: Find the equation of the line using $y - y_1 = m(x - x_1)$.

The perpendicular bisector has gradient 1 and passes through $(-2, 4)$. The equation of the perpendicular bisector is:

$$y - 4 = 1(x - (-2))$$
$$y - 4 = x + 2$$
$$y = x + 6$$

Example 3.12 The point A has coordinates $(3, -5)$ and the point B has coordinates $(5, 3)$. The midpoint of AB is M and the line MC is perpendicular to AB, where C has coordinates $(8, p)$.

  a Find the coordinates of M.

  b Find the gradient of MC.

  c Find the value of p.

Step 1: Draw a sketch.

Step 2: Use the midpoint formula.

  a Midpoint of $AB = \left(\frac{3 + 5}{2}, \frac{(-5) + 3}{2}\right) = (4, -1)$

Step 3: Use the gradient of the given line to find the gradient of the perpendicular line.

  b Gradient of $AB = \left(\frac{3 - (-5)}{5 - 3}\right) = \frac{8}{2} = 4$

Since AB and MC are perpendicular, the product of the gradients is -1. So the gradient of MC is $-\frac{1}{4}$.

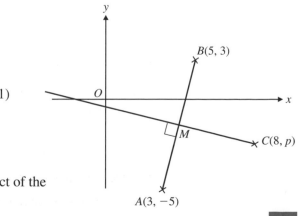

29

Step 4: Use the gradient formula for the given points.

c Using points $M(4, -1)$ and $C(8, p)$:

gradient of $MC = \dfrac{p - (-1)}{8 - 4} = \dfrac{p + 1}{4}$.

So $\dfrac{p + 1}{4} = -\dfrac{1}{4}$

$p + 1 = -1$

$p = -2$

Example 3.13 The line AB has equation $3x - 2y = 13$ and the points A and B have coordinates $(3, -2)$ and $(5, k)$ respectively.

  a Find the value of k.

  b Find the equation of the line through A that is perpendicular to AB, giving your answer in the form $ax + by + c = 0$, where a, b and c are integers.

Step 1: Substitute coordinates of B into equation of line.

a Since $(5, k)$ lies on the line $3x - 2y = 13$:

$3 \times 5 - 2k = 13$

$k = 1$

Tip:
If $(5, k)$ lies on the line, then $x = 5$ and $y = k$ must satisfy the equation of the line.

Step 2: Find the gradient of AB.

b A is the point $(3, -2)$, B is the point $(5, 1)$.

Gradient of $AB = \dfrac{1 - (-2)}{5 - 3} = \dfrac{3}{2}$

Tip:
Leave the gradient as a top-heavy fraction.

Step 3: Use the condition for gradients of perpendicular lines.

Gradient of the line perpendicular to $AB = -\dfrac{2}{3}$

Equation of line through $A(3, -2)$ with gradient $-\dfrac{2}{3}$ is given by:

Step 4: Use $y - y_1 = m(x - x_1)$.

$y - (-2) = -\dfrac{2}{3}(x - 3)$

$y + 2 = -\dfrac{2}{3}(x - 3)$

Step 5: Rearrange into required format.

$3(y + 2) = -2(x - 3)$

$3y + 6 = -2x + 6$

$2x + 3y = 0$

Tip:
Take care with fractions and negatives when rearranging the equation.

SKILLS CHECK **3B: Parallel and perpendicular lines**

1 State whether the following pairs of lines are parallel, perpendicular or neither:

  a $5y = 4x - 7$ ◉ **b** $y = \dfrac{2}{3}x - 8$ **c** $3x - 2y + 9 = 0$

      $4y = 7 - 5x$ $4x - 6y = 5$ $2x - 3y = 6$

2 Show that the following lines are perpendicular:
$3x - 4y = 20$ and $8x + 6y + 15 = 0$

3 Find the equation of the line parallel to the line $2x - 3y = 6$, passing through the point $(0, 3)$.

4 Find the equation of the line perpendicular to the line $y = \dfrac{4}{5}x - 2$, passing through the point $(0, -2)$.

◉ **5** Find the equation of the line perpendicular to $x + 5y - 10 = 0$, passing through the point $(3, -1)$.

6 Find the equation of the perpendicular bisector of AB, where $A(4, -5)$ and $B(-2, -3)$.

7 Find the equation of the line parallel to $y = -1\dfrac{2}{3}x - 2\dfrac{1}{6}$ passing through $(-2, -3)$.

8 Find the equation of the line perpendicular to the line $y = 5x - 8$ and passing through $(2, 2)$.

9 $A(1, 0)$, $B(3, 5)$ and $C(7, 3)$ are three vertices of a parallelogram $ABCD$. Find:

   a the gradient of AB **b** the equation of CD

   c the gradient of BC **d** the equation of AD

   e the coordinates of D.

SKILLS CHECK **3B EXTRA** is on the CD

3.4 The equation of a circle

Understand that the equation $(x - a)^2 + (y - b)^2 = r^2$ represents the circle with centre (a, b) and radius r.

The general equation of a circle with centre (a, b) and radius r is $(x - a)^2 + (y - b)^2 = r^2$.

If the circle has centre $(0, 0)$ the equation is $x^2 + y^2 = r^2$.

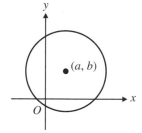

Note:
The brackets can be multiplied out to give the **expanded form** of the equation of a circle.

Example 3.14 **a** State the centre and radius of the circle
$$(x + 3)^2 + (y - 4)^2 - 25 = 0$$

b Write the equation of the circle in expanded form.

Step 1: Rearrange the equation into the general form.

a $(x + 3)^2 + (y - 4)^2 - 25 = 0$
$(x - (-3))^2 + (y - 4)^2 = 5^2$
The centre is at $(-3, 4)$ and the radius is 5.

Tip:
This is a translation of $x^2 + y^2 = 25$ by $\begin{pmatrix} -3 \\ 4 \end{pmatrix}$.
See Section 3.8.

Step 2: Expand and simplify.

b $(x + 3)^2 + (y - 4)^2 - 25 = 0$
$x^2 + 6x + 9 + y^2 - 8y + 16 - 25 = 0$
$x^2 + y^2 + 6x - 8y = 0$

Example 3.15 $A(3, -4)$ and $B(-5, 2)$ are the ends of a diameter of a circle. Find the equation of the circle.

Step 1: Draw a sketch.

Step 2: Find the centre of the circle.

The centre of the circle, C, is at the middle of the diameter.

$$\text{Midpoint of } AB = \left(\frac{3 + (-5)}{2}, \frac{-4 + 2}{2} \right) = (-1, -1)$$

so the centre of the circle is at $(-1, -1)$.

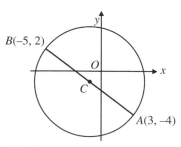

Step 3: Find the radius. $\text{Radius} = AC = \sqrt{(-1 - 3)^2 + (-1 - (-4))^2} = 5$

Tip:
Drawing a sketch will help you to see the situation and plan a strategy.

Step 4: Write the equation of the circle.

The equation of the circle is
$$(x - (-1))^2 + (y - (-1))^2 = 5^2$$
$$(x + 1)^2 + (y + 1)^2 = 25$$

Completing the square

Recall:
Completing the square (Section 2.5).

You will recall from work on quadratic functions that
$$x^2 + 2ax = (x + a)^2 - a^2.$$

You can use this to write the expanded form of the equation of a circle in the general form.

Example 3.16 A circle C has equation $x^2 + y^2 + 6x - 8y + 18 = 0$.

a By completing the square, express this equation in the form $(x - a)^2 + (y - b)^2 = r^2$.

b Write down the radius and the centre of the circle.

c Describe a geometrical transformation by which C can be obtained from the circle with equation $x^2 + y^2 = r^2$.

Step 1: Rewrite, grouping the x-terms and y-terms.
Step 2: Complete the square for both x and y.
Step 3: Collect all constant terms on the right-hand side.
Step 4: Compare with the general equation of a circle.
Step 5: Apply the translation.

a
$$x^2 + 6x + y^2 - 8y + 18 = 0$$
$$(x + 3)^2 - 9 + (y - 4)^2 - 16 + 18 = 0$$
$$(x + 3)^2 + (y - 4)^2 = 7$$

b The centre is at $(-3, 4)$ and the radius is $\sqrt{7}$.

c Translate O, the centre of the circle $x^2 + y^2 = 7$, three units to the left and four units up to obtain C, i.e. translate by the vector $\begin{pmatrix} -3 \\ 4 \end{pmatrix}$.

Recall:
Translations (Section 3.8).

Example 3.17 Find the centre and radius of the circle $x^2 + y^2 = 10y$ and show the circle on a sketch.

Step 1: Rewrite, grouping the x-terms and y-terms.

$$x^2 + y^2 = 10y$$
$$x^2 + y^2 - 10y = 0$$

Step 2: Complete the square for both x and y.

$$x^2 + (y - 5)^2 - 25 = 0$$

Step 3: Collect all constant terms on the right-hand side.

$$x^2 + (y - 5)^2 = 25$$

Hence the centre is at $(0, 5)$ and the radius is 5.

Step 4: Sketch the circle.

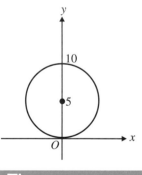

Tip:
Translate $x^2 + y^2 = 25$ by the vector $\begin{pmatrix} 0 \\ 5 \end{pmatrix}$.

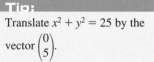

SKILLS CHECK **3C: The equation of a circle**

1 Give the general equation of the following circles.

a Centre $(3, -2)$, radius 4   **b** Centre $(-5, 0)$, radius 5

2 Write down the centre and radius of the following circles.

a $(x - 2)^2 + (y - 4)^2 = 6^2$ **b** $x^2 + (y + 3)^2 = 16$

3 Write the equation of the following circles in the expanded form.

a $(x + 1)^2 + (y - 2)^2 = 5^2$ **b** $(x + 3)^2 + (y - 4)^2 = 25$

c $(x - 2)^2 + (y - 6)^2 = 20$ **d** $(x + 2)^2 + (y - 2)^2 = 16$

4 Write the following circles in general form and hence state the centre and radius.

   a $x^2 + y^2 - 2x - 4y - 20 = 0$ **b** $x^2 + y^2 + 10x + 24y = 0$

   c $x^2 + y^2 - 6x + 10y + 18 = 0$ **d** $x^2 + y^2 + 2x - 6y + 3 = 0$

5 Find the equation of the circle with centre $C(2, -5)$ and point $A(6, -8)$ on the circumference.

6 Find the equation of the circle with diameter AB where A has coordinates $(4, -5)$ and B has coordinates $(-2, -3)$.

7 Find the centres and radii of the circles $x^2 + y^2 - 2x - 2y - 3 = 0$ and $x^2 + y^2 - 14x - 8y + 45 = 0$. Hence show that the circles touch each other.

8 The circle with equation $x^2 + y^2 - 4x - 8y + 7 = 0$ has centre C. The point $P(5, 2)$ lies on the circle.

   a Find the gradient of PC.

   b Find the equation of the line passing through P that is at right angles to the radius.

9 A circle C has equation $x^2 + y^2 = 4x - 6y + 3$.

   a Write the equation in the form $(x - a)^2 + (y - b)^2 = r^2$.

   b Write down the radius and the coordinates of the centre of the circle.

   c Describe a geometrical transformation by which C can be obtained from the circle with equation $x^2 + y^2 = r^2$.

10 A circle with equation $x^2 + y^2 = 49$ is translated 3 units down and 5 units to the right. Write down the equation of the translated curve in expanded form.

SKILLS CHECK **3C EXTRA** is on the CD

3.5 Coordinate geometry of a circle

Knowledge of the following circle properties: the angle in a semicircle is a right angle: the perpendicular from the centre to a chord bisects the chord; the perpendicularity of radius and tangent. Use algebraic methods to solve problems involving lines and circles, including the use of the equation of a circle in expanded form $x^2 + y^2 + 2gx + 2fy + c = 0$.

There are several circle properties that are useful in solving circle problems. Make sure that you can apply the following.

Circle property 1: angle in a semicircle

The angle subtended at the circumference of a circle by a diameter is a right angle.

Another way of saying this is: the angle in a semicircle is a right angle.

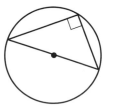

Example 3.18   **a** Show that the triangle ABC is right-angled, where A is $(0, -2)$, B is $(6, 6)$ and C is $(7, -1)$.

       b Given that A, B and C lie on the circumference of a circle, find the centre and radius of the circle and write down its equation.

Tip:
A fairly accurate sketch can be useful to help decide on a strategy to solve the problem. In this case it helps to identify where the right angle is.

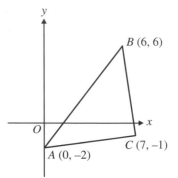

Step 1: Do a sketch.　**a**　Using gradient $= \dfrac{y_2 - y_1}{x_2 - x_1}$

Step 2: Show that two sides of the triangle are perpendicular.

Gradient $BC = \dfrac{-1 - 6}{7 - 6} = -7$

Gradient $AC = \dfrac{-1 - (-2)}{7 - 0} = \dfrac{1}{7}$

Since $\text{grad}_{BC} \times \text{grad}_{AC} = -1$, BC and AC are perpendicular and the triangle ABC is right-angled at C.

Step 3: Use circle properties to find the centre and radius.

b　The angle in a semicircle is a right angle, so AB is a diameter of the circle. The centre of the circle is the midpoint of AB, call this M. The radius is given by the length of AM.

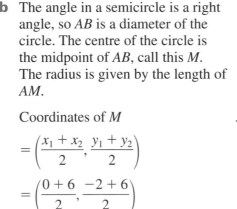

Note:
You could find AB^2, AC^2 and BC^2 and check Pythagoras' theorem.

Coordinates of M

$= \left(\dfrac{x_1 + x_2}{2}, \dfrac{y_1 + y_2}{2} \right)$

$= \left(\dfrac{0 + 6}{2}, \dfrac{-2 + 6}{2} \right)$

$= (3, 2)$

$AM = \sqrt{(3 - 0)^2 + (2 - (-2))^2} = 5$

The circle has centre $(3, 2)$ and radius 5.
The equation of the circle is $(x - 3)^2 + (y - 2)^2 = 25$.

Example 3.19　$A(-5, -1)$, $B(7, 3)$ and $C(-1, 7)$ lie on a circle with diameter AB and centre P.

a　Find the coordinates of P.

b　Find the equation of the perpendicular bisector of AC, writing your answer in the form $ax + by + c = 0$, where a, b and c are integers.

c　Show that the perpendicular bisector of AC passes through P.

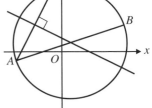

Step 1: Find the centre of the circle using circle properties.

a　AB is a diameter, so P is at the midpoint of AB.

Coordinates of $P = \left(\dfrac{-5 + 7}{2}, \dfrac{-1 + 3}{2} \right) = (1, 1)$

Step 2: Put the additional information on the sketch.

b　Gradient of $AC = \dfrac{7 - (-1)}{-1 - (-5)} = \dfrac{8}{4} = 2$

Step 3: Find the equation of the perpendicular bisector of AC.

Gradient of perpendicular bisector $= -\dfrac{1}{2}$

Midpoint of $AC = \left(\dfrac{-5 + (-1)}{2}, \dfrac{-1 + 7}{2} \right) = (-3, 3)$

Recall:
The product of the gradients of perpendicular lines is -1 (Section 3.3).

The perpendicular bisector has gradient $-\frac{1}{2}$ and goes through $(-3, 3)$. The equation of the perpendicular bisector of AC is

$$y - 3 = -\tfrac{1}{2}(x - (-3))$$
$$2(y - 3) = -(x + 3)$$
$$2y - 6 = -x - 3$$
$$x + 2y - 3 = 0$$

Step 4: Check that the coordinates of P satisfy the equation of the line.

c When $x = 1$ and $y = 1$:

$$x + 2y - 3 = 1 + 2 - 3 = 0$$

Therefore $(1, 1)$ lies on the line $x + 2y - 3 = 0$ so the perpendicular bisector of AC passes through P.

Recall:
The equation of the line through (x_1, y_1) with gradient m is $y - y_1 = m(x - x_1)$ (Section 3.2).

Circle property 2: perpendicular from the centre to a chord

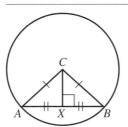

The perpendicular from the centre to a chord bisects the chord.

Note also:

- CAB is an isosceles triangle
- CX is the perpendicular bisector of chord AB.

Example 3.20 AB is a chord of a circle. A is the point $(-2, 1)$ and B is the point $(4, -1)$. The centre of the circle is $C(2, k)$.

a Find the midpoint M of AB.

b Find the gradient of AB.

c Find the equation of the line CM.

d Hence, or otherwise, find the value of k.

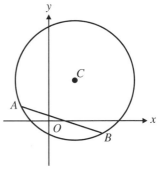

Step 1: Use the midpoint formula.

a $M = \left(\dfrac{-2 + 4}{2}, \dfrac{1 + (-1)}{2}\right) = (1, 0)$

Tip:
Show M on the sketch; see below.

Step 2: Use the gradient formula.

b Gradient of $AB = \dfrac{-1 - 1}{4 - (-2)} = -\dfrac{1}{3}$

Step 3: Find the equation of the perpendicular bisector of AB.

c $AM = MB \Rightarrow CM$ is the perpendicular bisector of AB.
Gradient of $CM = 3$

Equation of CM:
$$y - 0 = 3(x - 1)$$
$$y = 3x - 3$$

Tip:
Use $m_1 \times m_2 = -1$ for perpendicular lines.

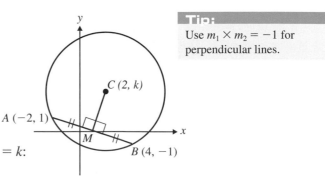

Step 4: Substitute point into the equation of the line.

d C lies on the line, so when $x = 2$ and $y = k$:
$$k = 3 \times 2 - 3 = 3$$
So $k = 3$.

Example 3.21 A circle with centre $C(1, 4)$ has chord AB where A is the point $(-4, 2)$ and B is the point $(3, k)$. Find the possible values of k.

Note:
At this stage you don't know exactly where B is. But you do know it lies on $x = 3$, so there are two possible positions.

Step 1: Draw a sketch.

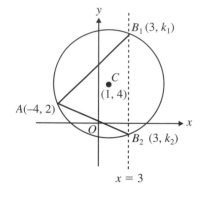

Step 2: Calculate the squares of the distances from C to A and B.

$CA^2 = (1 - (-4))^2 + (4 - 2)^2 = 29$

$CB^2 = (1 - 3)^2 + (4 - k)^2 = 4 + 16 - 8k + k^2 = 20 - 8k + k^2$

Step 3: Equate the distances and solve for k.

$CB = CA$ (both are radii) $\Rightarrow CB^2 = CA^2$

$20 - 8k + k^2 = 29$

$k^2 - 8k - 9 = 0$

$(k - 9)(k + 1) = 0$

$k = 9$ or $k = -1$

Recall:
Quadratic equations (Section 2.6).

Note:
The two values for k confirm the two possible positions for B of $(3, 9)$ and $(3, -1)$.

Circle property 3: tangent to a circle

The tangent to a circle is perpendicular to the radius at its point of contact.

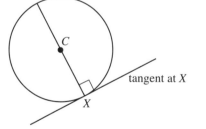

Example 3.22 $A(-1, 6)$ is a point on the circumference of a circle, centre $C(5, -3)$.
Find the gradient of the tangent at A.

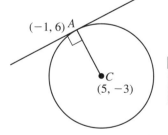

Step 1: Draw a sketch.
Step 2: Find the gradient of the radius at A.

Gradient $CA = \dfrac{-3 - 6}{5 - (-1)} = \dfrac{-9}{6} = -\dfrac{3}{2}$

Tip:
Leave gradients as top-heavy fractions.

Step 3: Find the gradient of the line perpendicular to the radius.

The tangent is perpendicular to the radius
$\Rightarrow$ the gradient of the tangent is $\frac{2}{3}$.

Example 3.23 A circle has centre $C(-3, 2)$.
The line $y = x - 1$ is a tangent to the circle at A.

a Find the gradient of CA.

b Find the equation of CA.

c Find the coordinates of A.

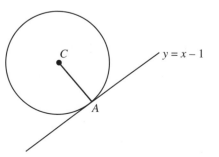

Step 1: Use circle properties.

a Gradient of tangent $= 1$

The radius is perpendicular to the tangent

$\Rightarrow$ gradient of $CA = -1$

Recall:
The gradient of the line $y = mx + c$ is m (Section 3.2).

Step 2: Use the equation of a line $y - y_1 = m(x - x_1)$.

b Equation of CA:

$y - 2 = -1(x - (-3))$

$y - 2 = -(x + 3)$

$y - 2 = -x - 3$

$\qquad y = -x - 1$

Recall:
The product of gradients of perpendicular lines is -1 (Section 3.3).

Step 3: Find the point of intersection of the lines by solving the simultaneous equations.

c A is the point of intersection of the lines

$y = x - 1 \qquad$ ①

$y = -x - 1 \qquad$ ②

Recall:
Solving two linear simultaneous equations (Section 2.7).

Substituting for y from ① into ②

$x - 1 = -x - 1$

$\quad 2x = 0$

$\quad\ x = 0$

Substituting for x in ①

$y = 0 - 1 = -1$

So the coordinates of A are $(0, -1)$.

Example 3.24 Find the equation of the tangent to the circle, centre $C(4, 3)$, at the point $A(5, -2)$ on the circumference. Write your answer in the form $ax + by + c = 0$, where a, b and c are integers.

Step 1: Draw a sketch.

Step 2: Find the gradient of the radius and use it to find the gradient of the tangent.

Gradient $CA = \dfrac{-2 - 3}{5 - 4} = -5$

Gradient of tangent $= \frac{1}{5}$

Equation of tangent at A:

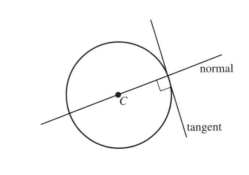

Step 3: Use $y - y_1 = m(x - x_1)$ to find the equation of the tangent.

$y - (-2) = \frac{1}{5}(x - 5)$

$5(y + 2) = x - 5$

$5y + 10 = x - 5$

$x - 5y - 15 = 0$

SKILLS CHECK **3D: Coordinate geometry of a circle**

1 $A(-4, 5)$, $B(4, 3)$ and $P(1, 0)$ are the vertices of triangle ABP.

a Show that angle $APB = 90°$.

b Hence show that P lies on the circumference of a circle with diameter AB.

c The circle has centre C. Find the coordinates of C and the radius of the circle.

d Hence write down the equation of the circle.

2 The point $(2, k)$ lies on the circumference of a circle with diameter AB, where A is the point $(-4, 1)$ and B is the point $(3, 2)$.
Find the possible values of k.

3 AB is the diameter of a circle, where $A(2, 2)$ and $B(p, q)$ lie on the line $x + 2y - 6 = 0$. $C(8, 2)$ lies on the circumference of the circle.

Tip:
Hint to **b**: AB is a diameter, C is on the circumference. What size is angle ACB?

   a Draw a sketch showing the line, A and C.

   b Find the equation of the line CB.

   c Find the values of p and q.

4 $A(1, -2)$ and $B(-5, 4)$ are the ends of a chord of a circle and C is the point $(-1, 2)$.

   a Show that triangle ACB is isosceles.

   b Could C be the centre of the circle? Give a reason for your answer.

5 $A(-3, 3)$ and $B(5, 1)$ are the two ends of a chord AB of a circle.

   a Find the midpoint of AB.

   b Show, by using gradients, that $C(0, -2)$ could be the centre of the circle.

6 AB and CD are two chords of a circle, where A is $(1, 5)$, B is $(5, 3)$, C is $(3, -1)$ and D is $(5, 1)$.

   a Find the equation of the perpendicular bisector of AB.

   b Find the equation of the perpendicular bisector of CD.

   c Hence find the coordinates of the centre of the circle.

7 $M(1, -3)$ is the midpoint of a chord AB of the circle $x^2 + y^2 = 20$. Find the equation of AB.

8 The point $P(5, 1)$ lies on a circle with centre $C(1, 4)$.

   a Show that P also lies on the line L with equation $3y - 4x + 17 = 0$.

   b Show that CP is perpendicular to L.

   c Deduce that L is a tangent to the circle at P.

9 Find the equation of the tangent to the circle, centre $C(3, 5)$, at the point $A(1, 3)$.

10 Find the equation of the tangent to the circle $x^2 + y^2 + 4x - 6y + 8 = 0$ at the point $A(-1, 1)$.

11 a A circle C has centre $(-3, 2)$. Find the equations of the tangents to the circle at the points $A(1, 1)$ and $B(-4, -2)$.

   b Hence find the point of intersection of the tangents.

12 $A(5, -1)$ and $B(7, 5)$ are two ends of a chord AB of a circle, centre $C(3, 3)$. Show that the tangents at A and B are perpendicular.

SKILLS CHECK **3D EXTRA** is on the CD

3.6 Geometrical interpretation of algebraic solution of equations

Understand the relationship between a graph and its associated algebraic equation, use points of intersection of graphs to solve equations and interpret geometrically the algebraic solution of equations (to include, in simple cases, understanding of the correspondence between a line being a tangent to a curve and a repeated root of an equation).

At a point where two curves, or a line and a curve, intersect or meet, the equation of each must be satisfied.

An algebraic way of finding the point of intersection is to solve the equations simultaneously.

Recall:
Simultaneous equations (Section 2.7).

The converse is also true:
If you solve two equations simultaneously you are effectively finding the point of intersection of the curves represented by the equations.

Example 3.25 The diagram shows a sketch of $y = x^2 - 9$ and $y = 4x - 12$.

The line and the curve intersect at A and B.

Find the coordinates of A and B.

Step 1: Solve the equations simultaneously.

When the line and curve intersect:

$$x^2 - 9 = 4x - 12$$
$$x^2 - 4x + 3 = 0$$
$$(x - 1)(x - 3) = 0$$
$$\Rightarrow \quad x = 1 \text{ or } x = 3$$

Substituting into $y = 4x - 12$:
When $x = 1$, $y = 4 \times 1 - 12 = -8$
When $x = 3$, $y = 4 \times 3 - 12 = 0$

Step 2: Use the solutions to write down the point(s) of intersection.

A is the point $(1, -8)$ and B is the point $(3, 0)$.

When finding the point(s) of intersection, you may have to solve a quadratic equation $ax^2 + bx + c = 0$. Here are the conditions relating to the discriminant of the quadratic polynomial $ax^2 + bx + c$. There are three cases to consider.

Recall:
The discriminant of a quadratic function (Section 2.3).

1 If $b^2 - 4ac > 0$ there are two real distinct roots. This means there are two points of intersection and the line cuts the curve twice.

2 If $b^2 - 4ac = 0$ there are two equal roots (repeated roots), so there is one point of intersection. The line touches the curve, so the line is a tangent to the curve at the point of intersection.

3 If $b^2 - 4ac < 0$ there are no real roots. The line and curve do not intersect or touch.

Example 3.26 Find the points of intersection of the circle $x^2 + y^2 = 20$ and the line $y = 3x - 10$.

Recall:
Solving simultaneous equations (Section 2.7).

Step 1: Substitute for *y* into the equation of the circle.

At the points of intersection, both these equations are satisfied.

$$y = 3x - 10 \qquad ①$$
$$x^2 + y^2 = 20 \qquad ②$$

Substituting for *y* from ① into ②:

$$x^2 + (3x - 10)^2 = 20$$

Step 2: Expand and rearrange into f(*x*) = 0.

$$x^2 + 9x^2 - 60x + 100 = 20$$
$$10x^2 - 60x + 80 = 0$$

Step 3: Solve the quadratic equation.

$$(\div 10) \quad x^2 - 6x + 8 = 0$$
$$(x - 4)(x - 2) = 0$$
$$x = 4 \text{ or } x = 2$$

Note:
There may be two answers and they will be coordinates: there will be an *x*-value and a *y*-value.

Note:
For the quadratic equation $x^2 - 6x + 8 = 0$, $a = 1$, $b = -6$, $c = 8$.
The discriminant:
$b^2 - 4ac = 36 - 32 > 0 \Rightarrow$ there are two points of intersection.

Step 4: Find the corresponding *y*-coordinate.

Substituting in ①:

when $x = 4$, $y = 3 \times 4 - 10 = 2$

when $x = 2$, $y = 3 \times 2 - 10 = -4$

The line and circle intersect at $(4, 2)$ and $(2, -4)$.

Example 3.27 Show that the line $y - x = -8$ and the circle $x^2 + y^2 - 6x - 8y = 0$ do not intersect.

Step 1: Rearrange the linear equation to make *x* or *y* the subject.

$$y - x = -8 \quad \Rightarrow y = x - 8$$

Step 2: Substitute for *y* into the equation of the circle.

Substituting for *y* in the equation of the circle:

$$x^2 + (x - 8)^2 - 6x - 8(x - 8) = 0$$
$$x^2 + x^2 - 16x + 64 - 6x - 8x + 64 = 0$$

Step 3: Expand and rearrange into f(*x*) = 0.

$$2x^2 - 30x + 128 = 0$$
$$(\div 2) \qquad x^2 - 15x + 64 = 0$$
$$a = 1, b = -15, c = 64$$

Step 4: Test the discriminant.

$$b^2 - 4ac = (-15)^2 - 4 \times 1 \times 64 = 225 - 256 = -31$$

Since $b^2 - 4ac < 0$ the equation has no real roots.

Therefore the line and the circle do not intersect.

Example 3.28   The line $y = 4x + k$ is a tangent to the curve $y = x^2 + 2x$. Find k.

Step 1: Substitute for y into the equation of the circle.

The line and curve touch when both these equations are satisfied.

$$y = 4x + k \qquad ①$$
$$y = x^2 + 2x \qquad ②$$

Substituting for y from ① into ②

Step 2: Expand and rearrange into $f(x) = 0$.

$$4x + k = x^2 + 2x$$
$$x^2 - 2x - k = 0$$

Step 3: Use the condition for the discriminant.

Since the line is a tangent, $x^2 - 2x - k = 0$ has two equal roots.

$$a = 1, b = -2, c = -k$$
$$b^2 - 4ac = 0 \Rightarrow (-2)^2 - 4 \times 1 \times (-k) = 0$$
$$4 + 4k = 0$$
$$k = -1$$

SKILLS CHECK   **3E: Intersection of lines and curves**

1 Find the point(s) of intersection of the following lines and curves.

   a $x^2 + y^2 = 25$ and $x + y = 7$

   b $y = x^2$ and $y = 4x - 4$

   c $x^2 + y^2 - 10x - 4y + 12 = 0$ and $y = x + 2$

   d $x^2 + y^2 - 4x - 2y - 15 = 0$ and $2x + y + 5 = 0$

2 Show that the line $y = x - 3$ is a tangent to the circle with centre $(-2, 1)$, radius $\sqrt{18}$, and find the point of contact.

3 Show that if the line $y = 2x - k$ does not intersect the curve $y = x^2 - 2$ then $k > 3$.

4 Solve the following pairs of simultaneous equations, if possible, and interpret each result geometrically.

   a $y = x^2 + 2x - 3$ and $y = 4x - 4$ **b** $(x - 3)^2 + (y + 4)^2 = 25$ and $y = 0$

   c $y = 2x^2$ and $y = 8x - 8$ **d** $x^2 + y^2 = 2x$ and $x + y = 3$

5 The line $y = x + k$ is a tangent to the curve $y = x^2 - x + 2$. Find the value of k.

6 The straight line $y = x + 7$ intersects the circle $x^2 + y^2 = 25$ at the points P and Q.

   a   Show that the x-coordinates of P and Q satisfy the equation $x^2 + 7x + 12 = 0$.

   b   Hence find the coordinates of P and Q.

7 A circle, centre $C(4, 3)$, touches the line with equation $3x - 2y + 7 = 0$ at P.

   a   Find the equation of CP.

   b   By solving the equations simultaneously, find the coordinates of P.

   c   Find the equation of the circle.

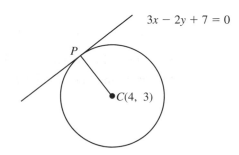

SKILLS CHECK   **3E EXTRA** is on the CD

Sketch curves with equations of the form

$y = kx^n$, where n is a positive or negative integer and k is a constant,

$y = k\sqrt{x}$, where k is a constant,

$y = ax^2 + bx + c$, where a, b and c are constants,

$y = f(x)$, where $f(x)$ is the product of at most three linear factors, not necessarily all distinct.

To sketch a curve, identify

- what general shape the curve takes
- where the curve crosses the x-axis, by setting $y = 0$
- where the curve crosses the y-axis, by setting $x = 0$.

> **Note:**
> You must indicate the intercepts with the axes on the sketch.

Graphs of linear functions

The graph of $y = kx$ is a straight line through $(0, 0)$ with gradient k.
For example,

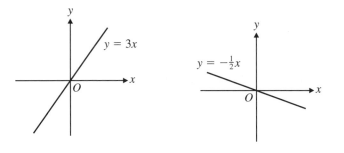

> **Recall:**
> See Section 3.2 for alternative formats.

The general equation of a straight line is $y = ax + b$, where a is the gradient and b is the y-intercept.

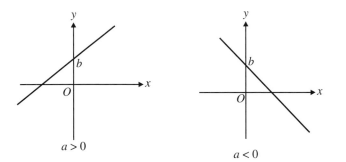

Graphs of quadratic functions

The graph of $y = kx^2$ is a parabola.

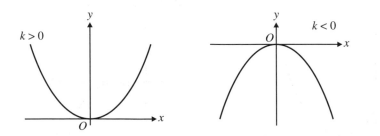

The general equation of a quadratic function is $y = ax^2 + bx + c$, where $a \neq 0$.

To find the intercepts with the axes:

$x = 0 \Rightarrow y = c$, so the curve passes through $(0, c)$.

$y = 0 \Rightarrow ax^2 + bx + c = 0$, so the intercepts on the x-axis are found by solving the quadratic equation.

Note that the axis of symmetry of the parabola is a vertical line passing through the midpoint between the intercepts on the x-axis.

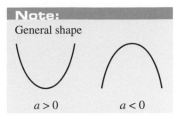

Note:
General shape

$a > 0$ $a < 0$

Example 3.29 It is given that $f(x) = 2x^2 + 5x - 3$.

  a   Sketch $y = f(x)$, labelling the intercepts with the axes.

  b   Draw the axis of symmetry on the sketch and state its equation.

Recall:
Quadratic equations (Section 2.6).

Recall:
Quadratic polynomials (Section 2.2).

Step 1: Decide the general shape.

  a $f(x)$ is a polynomial of degree 2, so the curve is a parabola.
$a > 0 \Rightarrow$ parabola is $\cup$-shaped.

Step 2: Set $y = 0$ and $x = 0$ to find the axes intercepts.

When $x = 0$, $y = -3$, so the curve goes through $(0, -3)$.

When $y = 0$,
$$2x^2 + 5x - 3 = 0$$
$$(2x - 1)(x + 3) = 0$$
$$\Rightarrow \quad x = 0.5 \text{ or } x = -3$$

Note:
The axis of symmetry and the vertex can also be found by completing the square (Section 2.5) or by differentiating (Section 4.6).

Step 3: Sketch the curve, marking the intercepts.

$(0.5, 0)$ and $(-3, 0)$ lie on the curve.

$$\text{Midpoint of } x\text{-intercepts} = \frac{-3 + 0.5}{2} = -1.25$$

Step 4: Draw the axis of symmetry.

  b   The axis of symmetry is the line $x = -1.25$.

Step 5: Find the axis of symmetry using the x-axis intercepts.

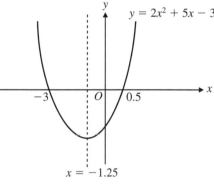

Significance of the discriminant

The discriminant can be used to determine whether or not the quadratic curve crosses or touches the x-axis.

If $b^2 - 4ac > 0$, the equation $ax^2 + bx + c = 0$ has two real distinct (different) roots. These give the two x-coordinates of the points where the graph of $y = ax^2 + bx + c$ crosses the x-axis.

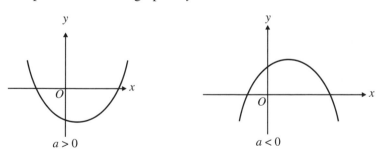

$a > 0$ $a < 0$

If $b^2 - 4ac = 0$, then $ax^2 + bx + c = 0$ has two equal roots (repeated roots). This gives the x-coordinate where the graph of $y = ax^2 + bx + c$ touches the x-axis.

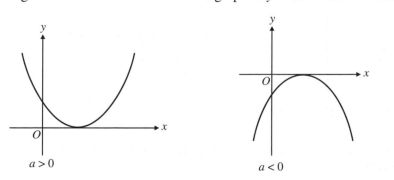

$a > 0$ $a < 0$

Note:
The x-axis is a tangent to the curve (see Sections 3.6 and 4.4).

If $b^2 - 4ac < 0$, then $ax^2 + bx + c = 0$ has no real roots.

This tells you that the graph of $y = ax^2 + bx + c$ does not cross or touch the x-axis.

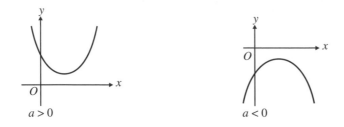

$a > 0$ $a < 0$

Example 3.30   Show that the curve $y = 3x^2 - 2x + 5$ lies entirely above the x-axis.

When $x = 0$, $y = 5$, so the curve passes through $(0, 5)$.

When $y = 0$, $3x^2 - 2x + 5 = 0$

Step 1: Find the discriminant. $a = 3$, $b = -2$, $c = 5$

$b^2 - 4ac = (-2)^2 - 4 \times 3 \times 5 = 4 - 60 = -56$

Step 2: Use condition on discriminant.   Since $b^2 - 4ac < 0$, the equation $3x^2 - 2x + 5 = 0$ has no real roots and the curve does not cross or touch the x-axis.

Since the curve passes through $(0, 5)$, it must lie entirely above the x-axis.

> **Note:**
> You could find the vertex by completing the square (Section 2.5) or using differentiation (Section 4.6).

Graphs of cubic functions

The diagrams show the graph of $y = kx^3$.

> **Note:**
> The highest power of x is 3.

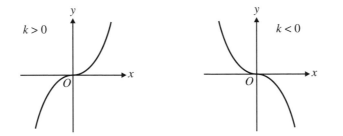

$k > 0$ $k < 0$

The general equation of a cubic curve is $y = ax^3 + bx^2 + cx + d$, where $a \neq 0$.

Intercepts with the axes:

$x = 0 \Rightarrow y = d$, so the curve passes through $(0, d)$.

$y = 0 \Rightarrow ax^3 + bx^2 + cx + d = 0$, so to find the intercepts on the x-axis, solve the cubic equation.

In general, the graph of $y = ax^3 + bx^2 + cx + d$ can have the following shapes:

$a > 0$

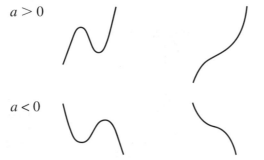

$a < 0$

> **Note:**
> A cubic curve may have two, one or no stationary points. See Section 4.6 for more on stationary points.

Example 3.31   Sketch the curve $y = (x - 1)(2x + 1)(2 - x)$.

Step 1: Decide the general shape.

The highest power of x is 3, so this is a cubic curve.

Coefficient of x^3 is -2, so the general shape is one of those described above for $a < 0$.

Tip:
Multiplying the terms indicated gives the coefficient of x^3:
$(x - 1)(2x + 1)(2 - x)$.

Step 2: Set $y = 0$ and $x = 0$ to find the axes intercepts.

When $y = 0$, $(x - 1)(2x + 1)(2 - x) = 0$

$\Rightarrow$ $x = 1, x = -\frac{1}{2}$ or $x = 2$

When $x = 0$, $y = (-1) \times 1 \times 2 = -2$

Step 3: Sketch the curve, marking the intercepts.

The curve goes through $(0, -2)$, $(-\frac{1}{2}, 0)$, $(1, 0)$ and $(2, 0)$.

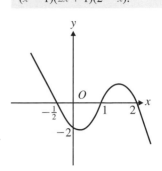

Graphs of reciprocal functions

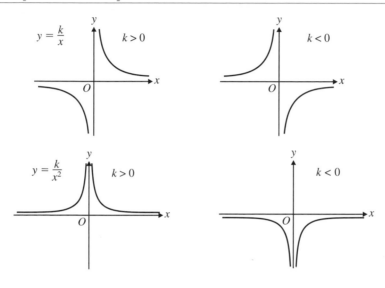

Note:
The x-axis and the y-axis are asymptotes

Graphs for general cases of $y = kx^n$ (where n is an integer)

In general, the graph of $y = kx^n$, where n is an even positive integer, has the same general shape as $y = kx^2$. The graph of $y = kx^n$ when n is an odd positive integer has the same general shape as $y = kx^3$.

Similarly, the graph $y = kx^n$, where n is a negative odd integer, has the same general shape as $y = kx^{-1} = \dfrac{k}{x}$. The graph of $y = kx^n$ when n is a negative even integer has the same general shape as $y = kx^{-2} = \dfrac{k}{x^2}$.

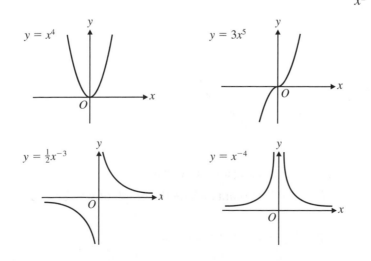

Tip:
Graphs with even powers of x are symmetrical about the y-axis and are in the first and second quadrants when $k > 0$.

Tip:
Graphs with odd powers of x have rotational symmetry about the origin and are in the first and third quadrants when $k > 0$.

Graph of $y = k\sqrt{x}$

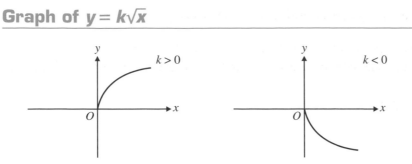

Intersecting curves

Example 3.32   Sketch the graphs of $y = x^{-1}$ and $y = x^4$ on the same axes and write down the coordinates of their point of intersection.

Step 1: Draw a pair of axes.

Step 2: Sketch the graphs.

Step 3: Solve the equations simultaneously.

The curves intersect when

$$x^{-1} = x^4 \Rightarrow \frac{1}{x} = x^4$$

$$x^5 = 1$$

$$x = \sqrt[5]{1} = 1$$

When $x = 1$, $y = 1$, so the point of intersection is $(1, 1)$.

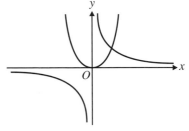

SKILLS CHECK   **3F: Sketching curves**

1 Sketch the following, showing the intercepts with the axes.

   a $y = 3x + 2$ **b** $y = 1 - 2x$ **c** $y = (x - 3)(2x + 1)$

   d $y = (3 - x)(2 + x)$ **e** $y = x^2 + 4x - 5$ **f** $y = 7 - 5x + 2x^2$

2 **a** Factorise $-2x^2 - 7x + 4$.

   b Sketch the graph of $y = -2x^2 - 7x + 4$, showing the intercepts with the axes.

3 Sketch the following graphs, showing the vertex and the coordinates of any intercepts with the axes.

   a $y = (x - 3)^2$ **b** $y = (3 - x)^2$

4 It is given that $f(x) = x^2 + 2x - 24$.

   a Factorise $f(x)$ and hence solve $f(x) = 0$.

   b Write $f(x)$ in the form $(x + B)^2 + C$.

   c Using your answers from parts **a** and **b**, sketch $y = f(x)$, showing the intercepts with the axes and the vertex.

   d Write down the equation of the axis of symmetry.

5 Sketch the following cubic graphs, showing the intercepts with the axes.

   a $y = (x + 3)(x + 4)(x - 2)$ **b** $y = (x + 1)(6 - x)(2 - 3x)$ **c** $y = x(x - 2)^2$

6 It is given that $f(x) = 2x^3 - x^2 - 5x - 2$.

   a By expanding the right-hand side, show that $2x^3 - x^2 - 5x - 2 = (2x + 1)(x + 1)(x - 2)$.

   b Find the coordinates of the points where the graph of $y = 2x^3 - x^2 - 5x - 2$ meets the coordinate axes.

   c Sketch the graph of $y = 2x^3 - x^2 - 5x - 2$.

7 Sketch the graphs of $y = 8\sqrt{x}$ and $y = x^2$ on the same axes. Find the points of intersection of the curves.

8 Sketch the graphs of $y = \frac{1}{2}x^{-1}$ and $y = 8x^3$ on the same axes.
Given that one point of intersection is $(\frac{1}{2}, 1)$, write down the other point of intersection.

SKILLS CHECK **3F EXTRA** is on the CD

3.8 Transformations

Understand and use the relationships between the graphs of $y = f(x)$, $y = a\,f(x)$, $y = f(x) + a$, $y = f(x + a)$, where a is a constant, and express the transformations involved in terms of translations, reflections and sketches.

Translations

$y = f(x) + a$

The transformation $y = f(x) + a$ has the effect of **translating** the graph of $y = f(x)$ by a units in the y-direction.

The vector form of the translation is $\begin{pmatrix} 0 \\ a \end{pmatrix}$.

If $a > 0$, the graph moves up.
If $a < 0$, the graph moves down.

For example:
The curve $y = x^2 + 2$ is a translation of the curve $y = x^2$ by 2 units up.

The curve $y = x^2 - 1$ is a translation of the curve $y = x^2$ by 1 unit down.

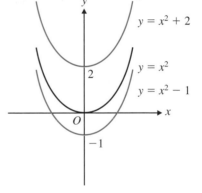

$y = f(x + a)$

The transformation $y = f(x + a)$ has the effect of **translating** the graph of $y = f(x)$ by $-a$ units in the y-direction.

The vector form of the translation is $\begin{pmatrix} -a \\ 0 \end{pmatrix}$.

If $a > 0$, the graph moves to the left.
If $a < 0$, the graph moves to the right.

For example:

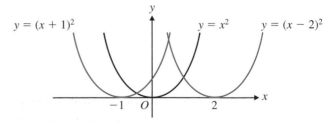

The curve $y = (x - 2)^2$ is a translation of the curve $y = x^2$ by 2 units to the right.

The curve $y = (x + 1)^2$ is a translation of the curve $y = x^2$ by 1 unit to the left.

Although you may be examined on the effect of one transformation on a given curve, it is useful to more the effect of performing two translations, one in the x-direction and one in the y-direction.

This is especially useful when sketching quadratic curves when the equation of the curve has been written in 'completed square' form, $y = (x + a)^2 + b$. This represents a translation of $y = x^2$ by $-a$ units in the x-direction and b units in the y-direction.

Note:
The vertex moves from $(0, 0)$ to $(-a, b)$.

47

This is illustrated in the following example.

Example 3.33
a Express $x^2 - 4x - 1$ in the form $(x + a)^2 + b$.

b State the transformation that maps $y = x^2$ onto $y = x^2 - 4x - 1$.

c Hence, or otherwise, sketch the graph of $y = x^2 - 4x - 1$, stating the coordinates of P, the minimum point on the curve.

Step 1: Complete the square.

a $x^2 - 4x - 1 = (x - 2)^2 - 4 - 1 = (x - 2)^2 - 5$

Step 2: Compare with $f(x + a) + b$.

b $y = x^2 - 4x - 1$ is a translation of $y = x^2$ by 2 units to the right and 5 units down.

Tip:
The negative of the value inside the bracket gives the x-translation and the constant term gives the y-translation.

Step 3: Sketch the curve and write down the coordinates of the minimum point.

c The vertex moves from $(0, 0)$ to $(2, -5)$, so the coordinates of P are $(2, -5)$.

Recall:
The minimum point on the curve $y = (x + a)^2 + b$ is $(-a, b)$ (Section 2.5).

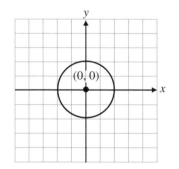

Translations can also be applied to circles. In general, the circle $(x - a)^2 + (y - b)^2 = r^2$ is a translation of the circle $x^2 + y^2 = r^2$ by $\begin{pmatrix} a \\ b \end{pmatrix}$.

Example 3.34 The sketch shows the circle $x^2 + y^2 = 4$.

a Sketch the circle $(x - 2)^2 + (y + 1)^2 = 4$

b State the coordinates of its centre.

Step 1: Identify the translation.

a Translate by $\begin{pmatrix} 2 \\ -1 \end{pmatrix}$.

Step 2: Sketch the graph in the new position.

b The centre of the circle is at $(2, -1)$.

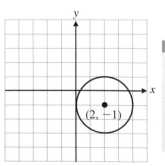

Note:
The effect of $(x - 2)^2$ is to translate the circle by $\begin{pmatrix} 2 \\ 0 \end{pmatrix}$. The effect of $(y + 1)^2$ is to translate the circle by $\begin{pmatrix} 0 \\ -1 \end{pmatrix}$.

Stretches

$y = af(x)$

The transformation $y = a f(x)$ has the effect of **stretching** the graph of $y = f(x)$ by a factor of a units in the y-direction. Points on the x-axis are invariant.

Note:
Invariant points do not move under the transformation.

For example, consider $y = x^3$ and $y = 2x^3$.

x	-2	-1	0	1	2
$y = x^3$	-8	-1	0	1	8
$y = 2x^3$	-16	-2	0	2	16

Note:
You can see from the table that, for $y = 2x^3$, all the y-coordinates are multiplied by 2.

The effect is to stretch the curve $y = x^3$ by 2 units in the y-direction.

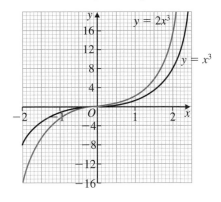

$y = f(ax)$

The transformation $y = f(ax)$ has the effect of **stretching** the graph of $y = f(x)$ by a factor of $\dfrac{1}{a}$ units in the x-direction. Points on the y-axis are invariant.

When $a > 1$, the graph appears more squashed in the x-direction. For $0 < a < 1$, the graph appears to be lengthened in the x-direction.

It is important to remember that the correct description in both cases is a *stretch* in the x-direction.

For example, the diagram shows a sketch of $y = f(x)$. A is the point $(-4, 0)$, B is the point $(-1, 5)$ and C is the point $(2, 0)$.

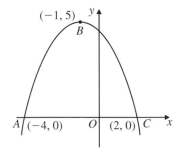

To obtain the graph of $y = f(2x)$, divide all the x-coordinates by 2 while keeping the same y-coordinates.
A moves to $(-2, 0)$, B moves to $(-0.5, 5)$ and C moves to $(1, 0)$.

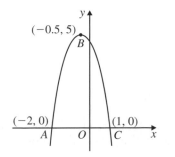

Note:
This is called a stretch, even though it looks squashed up.

Reflections

y = −f(x)

The transformation $y = -f(x)$ has the effect of reflecting the graph of $y = f(x)$ in the x-axis. For example, in the graph of $y = -(x + 2)$, all the points in the graph of $y = x + 2$ have been reflected in the x-axis.

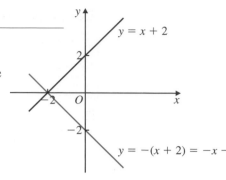

y = f(−x)

The transformation $y = f(-x)$ has the effect of reflecting the graph of $y = f(x)$ in the y-axis. For example, in the graph of $y = -x + 2$, all the points in the graph of $y = x + 2$ have been reflected in the y-axis.

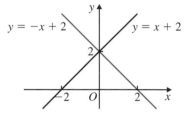

Mixed examples

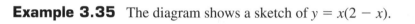

Example 3.35 The diagram shows a sketch of $y = x(2 - x)$.

The vertex P is the point $(1, 1)$.

By applying appropriate transformations, sketch the following curves:

a $y = 3x(2 - x)$ **b** $y = x(2 - x) - 1$

In each case describe the transformation, and state the coordinates of P_1, the vertex of the curve.

Step 1: Identify the transformation.

Step 2: Sketch the curve, applying the transformation.

Step 3: Describe the transformation and state the coordinates of the vertex.

a

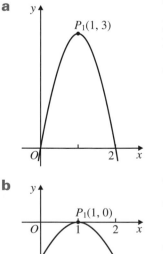

The curve $y = 3x(2 - x)$ is a stretch in the y-direction, factor 3, of the curve $y = x(2 - x)$.

The coordinates of the vertex P_1 are $(1, 3)$.

Tip:
The y-coordinates are multiplied by 3.

b

The curve $y = x(2 - x) - 1$ is a translation, by $\begin{pmatrix} 0 \\ -1 \end{pmatrix}$, of the curve $y = x(2 - x)$.

The coordinates of the vertex P_1 are $(1, 0)$.

Tip:
All points move 1 unit down.

Example 3.36 The diagram shows a sketch of $y = f(x)$ for $0 \leqslant x \leqslant 3$. For all other values of x $f(x) = 0$.

A is the point $(2, 1)$ and B is the point $(3, 0)$.

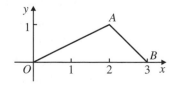

Describe the transformation and draw sketches to show

a $y = f(2x)$

b $y = f(\frac{1}{3}x)$

In each case, state the coordinates of the new positions of A and B.

Step 1: Identify the
transformation.

Step 2: Draw the sketch by
applying the
transformation.

Step 3: State the new
coordinates.

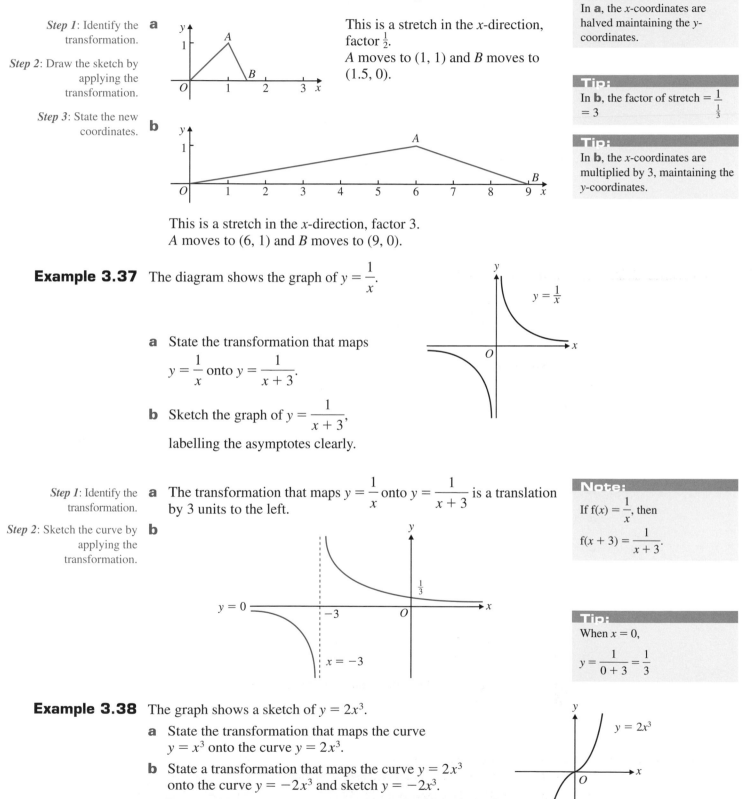

a This is a stretch in the x-direction, factor $\frac{1}{2}$.
A moves to $(1, 1)$ and B moves to $(1.5, 0)$.

Tip:
In **a**, the x-coordinates are halved maintaining the y-coordinates.

Tip:
In **b**, the factor of stretch $= \dfrac{1}{\frac{1}{3}} = 3$

Tip:
In **b**, the x-coordinates are multiplied by 3, maintaining the y-coordinates.

b This is a stretch in the x-direction, factor 3.
A moves to $(6, 1)$ and B moves to $(9, 0)$.

Example 3.37 The diagram shows the graph of $y = \dfrac{1}{x}$.

a State the transformation that maps
$y = \dfrac{1}{x}$ onto $y = \dfrac{1}{x + 3}$.

b Sketch the graph of $y = \dfrac{1}{x + 3}$,
labelling the asymptotes clearly.

Step 1: Identify the
transformation.

Step 2: Sketch the curve by
applying the
transformation.

a The transformation that maps $y = \dfrac{1}{x}$ onto $y = \dfrac{1}{x + 3}$ is a translation by 3 units to the left.

Note:
If $f(x) = \dfrac{1}{x}$, then
$f(x + 3) = \dfrac{1}{x + 3}$.

b

Tip:
When $x = 0$,
$y = \dfrac{1}{0 + 3} = \dfrac{1}{3}$

Example 3.38 The graph shows a sketch of $y = 2x^3$.

a State the transformation that maps the curve $y = x^3$ onto the curve $y = 2x^3$.

b State a transformation that maps the curve $y = 2x^3$ onto the curve $y = -2x^3$ and sketch $y = -2x^3$.

c State another transformation that maps the curve $y = 2x^3$ onto the curve $y = -2x^3$.

Step 1: Apply the transformation $y = a\,\mathrm{f}(x)$.

a The transformation that maps $y = x^3$ onto $y = 2x^3$ is a stretch of factor 2 in the y-direction.

Step 1: Apply the transformation $y = -\mathrm{f}(x)$.

b To map $y = 2x^3$ onto $y = -2x^3$ reflect in the x-axis.

Step 2: Sketch the curve.

Step 1: Apply the transformation $y = \mathrm{f}(-x)$.

c The curve $y = 2x^3$ can be mapped onto $y = -2x^3$ by reflecting in the y-axis.

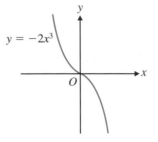

$y = -2x^3$

SKILLS CHECK **3G: Transformations**

1 For each of the following curves,

   i describe the geometrical transformation by which the curve can be obtained from the parabola with equation $y = x^2$,

   ii sketch the curve, stating the coordinates of the vertex and the y-intercept.

  a $y = x^2 + 3$ **b** $y = x^2 - 2$ **c** $y = x^2 + 1$

  d $y = (x + 2)^2$ **e** $y = (x - 1)^2$ **f** $y = (x + 4)^2$

  g $y = (x + 3)^2 + 1$ **h** $y = (x - 2)^2 - 1$ ⊚ **i** $y = (x + 1)^2 - 2$

2 For each of the following circles,

   i describe the geometrical transformation by which the circle can be obtained from the circle with equation $x^2 + y^2 = r^2$,

   ii state the coordinates of the centre of the circle and the radius.

  a $(x + 2)^2 + y^2 = 4$ **b** $x^2 + (y - 2)^2 = 25$

  c $(x + 1)^2 + (y - 3)^2 = 9$ **d** $(x - 2)^2 + (y + 1)^2 = 4$

3 State the equations of each of the following circles.

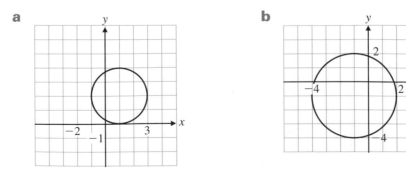

4 a Describe the transformation that maps $y = x^2$ onto

   i $y = x^2 + 2$ **ii** $y = -x^2$.

  b Sketch the three curves on the same axes.

5 a Describe the transformation that maps $y = x^3$ onto

    i $y = 2x^3$ **ii** $y = -x^3$.

  b Sketch the three curves on the same axes.

6 Describe the transformation that maps $y = \dfrac{1}{x}$ onto

  a $y = \dfrac{2}{x}$ **b** $y = \dfrac{1}{2x}$ **c** $y = \dfrac{1}{x + 1}$

7 The diagram shows the graph $y = f(x)$ for $0 \leqslant x \leqslant 2$.
The vertex A is the point $(1, 3)$.

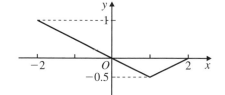

  a Sketch the graph of $y = -f(x)$ for $0 \leqslant x \leqslant 2$ and state the coordinates of the vertex of the curve $y = -f(x)$.

  b **i** Describe fully the transformation that maps $y = f(x)$ onto $y = 2f(x)$.

      ii Sketch the graph of $y = 2f(x)$ for $0 \leqslant x \leqslant 2$ and state the coordinates of the vertex of the curve $y = 2f(x)$.

8 The equation $y = x^2 + 2x - 3$ can be written as $y = (x + 1)^2 - 4$.
Describe the transformation that maps $y = x^2$ to $y = x^2 + 2x - 3$.

9 a Given that $f(x) = x^2$, state the transformation that maps $y = f(x)$ onto

    i $y = 4f(x)$ **ii** $y = f(2x)$

  b State the equation of each of the transformed curves and comment.

10 The sketch of $y = f(x)$ for $-2 \leqslant x \leqslant 2$ is shown in the diagram.

On separate axes, for $-2 \leqslant x \leqslant 2$, sketch the graphs of

  a $y = f(-x)$

  b $y = -f(x)$

  c $y = f(x) - 1$

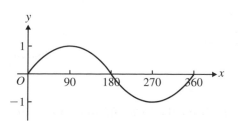

11 Which of the following transformations of the graph $y = x^2$ give the same graph?

  a Stretch factor 3 in the x-direction

  b Stretch factor 3 in the y-direction

  c Stretch factor 9 in the x-direction

  d Stretch factor 9 in the y-direction

12 a Given that $f(x) = x^3$, state the transformation that maps $y = f(x)$ onto

    i $y = -f(x)$ **ii** $f(-x)$

  b State the equation of each of the transformed curves and comment.

13 The graph shows $y = \sin x°$ for $0 \leqslant x \leqslant 360$.

Describe each of the following as transformations of $y = \sin x$ and write down the equation of each graph.

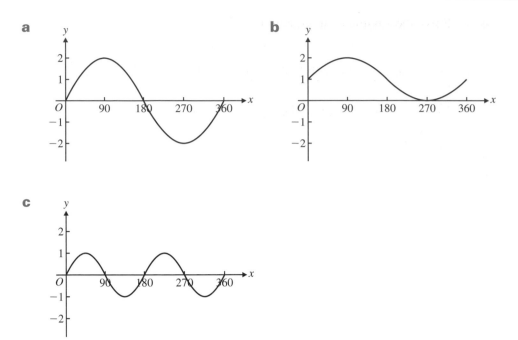

a

b

c

SKILLS CHECK **3G EXTRA** is on the CD

Examination practice Coordinate geometry and graphs

1 i Given that $f(x) = x^2$, sketch the graph of $y = f(x)$.

The graph of $y = g(x)$ is obtained by reflecting the graph of $y = f(x)$ in the x-axis.

The graph of $y = h(x)$ is obtained by translating the graph of $y = g(x)$ by $+2$ units parallel to the y-axis.

ii Sketch and label the graphs of $y = g(x)$ and $y = h(x)$ on a single diagram.

iii Write down expressions for $g(x)$ and $h(x)$ in terms of x.   [OCR May 2004]

2 i Sketch the graph of $y = 3\sqrt{x}$, for $x \geqslant 0$.

ii The graph of $y = 3\sqrt{x}$ is stretched by a factor of 2 parallel to the y-axis. State the equation of the transformed graph.

iii Describe the single geometrical transformation that transforms the graph of $y = 3\sqrt{x}$ to the graph of $y = 3\sqrt{(x - k)}$.   [OCR Jan 2003]

3 The equation $x^2 + 4kx + 3k = 0$, where k is a constant, has distinct real roots.

i Prove that $k(4k - 3) > 0$.

ii Hence find the set of possible values of k.

It is given instead that the x-axis is a tangent to the graph of $y = x^2 + 4kx + 3k$.

iii Write down the possible values of k.   [OCR Jan 2003]

4 i Solve the simultaneous equations
$$y = x^2 - 3x + 2, \quad y = 3x - 7.$$

ii What can you deduce from the solution to part **i** about the graphs of $y = x^2 - 3x + 2$ and $y = 3x - 7$?

iii Hence, or otherwise, find the equation of the normal to the curve $y = x^2 - 3x + 2$ at the point $(3, 2)$, giving your answer in the form $ax + by + c = 0$ where a, b and c are integers.   [OCR Spec paper]

5

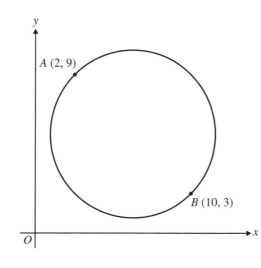

The diagram shows a circle which passes through the points $A(2, 9)$ and $B(10, 3)$. AB is a diameter of the circle.

i Calculate the radius of the circle and the coordinates of the centre.

ii Show that the equation of the circle may be written in the form $x^2 + y^2 - 12x - 12y + 47 = 0$.

iii The tangent to the circle at the point B cuts the x-axis at C. Find the coordinates of C.   [OCR Spec paper]

6 i Calculate the discriminant of $3x^2 + 5x + 8$.

ii Write down the number of points of intersection of the curve $y = 3x^2 + 5x + 8$ with the x-axis.

iii Show that $y = 11x + 5$ is a tangent to the curve $y = 3x^2 + 5x + 8$.   [OCR May 2003]

7 $ABCD$ is a rectangle, where A, B and C are the points $(3, 4)$, $(1, k)$ and $(4, -3)$ respectively.

i Find the coordinates of the midpoint of AC.

ii Find the gradient of the line AB, giving your answer in terms of k.

iii Determine the two possible values of k.

iv Find the area of the rectangle $ABCD$ for the case in which k is positive.   [OCR Nov 2003]

8 The points A, B and C have coordinates $(-3, 5)$, $(9, -3)$ and $(7, 7)$ respectively.

   i Find the equation of the straight line through A and B, giving your answer in the form $ax + by = c$.

  D is the midpoint of AB.

   ii Show that CD is perpendicular to AB.

   iii The line through A and B is a tangent to a circle with centre C. Find the area of the circle, giving your answer in terms of π.   [OCR May 2004]

9 The coordinates of the points A, B and C are $(-2, 3)$, $(2, 5)$ and $(4, 1)$ respectively.

   i Find the gradients of the lines AB, BC and CA.

   ii Hence, or otherwise, show that triangle ABC is a right-angled triangle.   [OCR Jan 2003]

10 The points A, B and C have coordinates $(1, 2)$, $(3, 4)$ and $(9, -2)$ respectively.

   i Calculate the gradient of AB.

   ii Show that BC is perpendicular to AB.

   iii Find an equation of the straight line passing through B and C.

   iv The length of AB may be written in the form $p\sqrt{2}$. Find the constant p.

   v Find the area of triangle ABC.   [OCR May 2003]

11 Find the midpoint of the line segment joining $(2, 4)$ and $(-3, 1)$.   [OCR Jan 2004]

4 Differentiation

4.1 Differentiation, gradients and rates of change

Understand the gradient of a curve at a point as the limit of the gradients of a suitable sequence of chords.

On the curve $y = f(x)$, the slope may vary for different values of x.

At a point P on the curve, a measure of the slope is the **gradient** at P. This tells you how y is changing in relation to x at that point. This is described as the **rate of change** of y with respect to x.

The **gradient of the curve** at P is defined as the **gradient of the tangent** at P. The tangent is the limiting position of the line through P and point Q on the curve as Q gets closer and closer to P.

Note:
The tangent at P *touches* the curve at P.

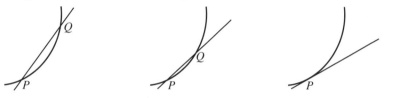

The gradient of the tangent at a point can be found by **differentiating** $y = f(x)$.

Tip:
See tangents and normals (Section 4.4).

This gives the **gradient function**, also known as the **derived function** or the **derivative** with respect to x.

The gradient function is usually written $\dfrac{dy}{dx}$ or $f'(x)$.

Note:
$\dfrac{dy}{dx}$ is read as 'dee y by dee x' and $f'(x)$ is read as 'f dashed x'.

4.2 Differentiation of x^n

Understand the ideas of a derived function and second order derivative, and use the notations $f'(x)$, $f''(x)$, $\dfrac{dy}{dx}$ and $\dfrac{d^2y}{dx^2}$. Use the derivatives of x^n (for any rational n), together with constant multiples, sums and differences.

In Core 1 you need to be able to differentiate positive powers of x, using the rule:

$$y = x^n \Rightarrow \frac{dy}{dx} = nx^{n-1}$$

Tip:
Multiply by the power of x and decrease the power by 1.

If a is a constant:

$$y = ax^n \Rightarrow \frac{dy}{dx} = nax^{n-1}$$

Note:
In function notation, $f(x) = ax^n$ $\Rightarrow f'(x) = nax^{n-1}$.

It is useful to remember the following:

$$y = ax \Rightarrow \frac{dy}{dx} = a$$

$$y = a \Rightarrow \frac{dy}{dx} = 0$$

Note:
$y = ax$ is a line, with constant gradient.
$y = a$ is a line parallel to the x-axis, with zero gradient.

Example 4.1 Find $\dfrac{dy}{dx}$ when   **a** $y = x^3$ **b** $y = 4x^2$ **c** $y = 5$.

Step 1: Use the differentiation rule.

a $\quad y = x^3$

$$\frac{dy}{dx} = 3x^2$$

b $\quad y = 4x^2$

$$\frac{dy}{dx} = 2 \times 4x^1 = 8x$$

c $\quad y = 5$

$$\frac{dy}{dx} = 0$$

To differentiate an expression in x containing several terms, for example $3x^7 - 2x^3 + 3x - 1$, differentiate the terms individually, using the rule:

$$y = f(x) \pm g(x) \Rightarrow \frac{dy}{dx} = f'(x) \pm g'(x)$$

Example 4.2 It is given that $y = 3x^7 - 2x^3 + 3x - 1$. Find the derivative of y with respect to x.

Step 1: Differentiate each term separately.

$$y = 3x^7 - 2x^3 + 3x - 1$$
$$\frac{dy}{dx} = 21x^6 - 6x^2 + 3$$

Example 4.3 Find $f'(x)$ where $f(x) = (2x - 3)^2$.

Step 1: Expand. $f(x) = (2x - 3)^2 = 4x^2 - 12x + 9$

Step 2: Differentiate term by term. $f'(x) = 8x - 12$

Example 4.4 Differentiate with respect to x:

a $y = \dfrac{1}{x^2}$

b $y = 3\sqrt{x} + \dfrac{2}{x^4}$

Step 1: Write the terms in index form, ax^n.

a $y = \dfrac{1}{x^2} = x^{-2}$

Step 2: Differentiate term by term using the rule.

$$\frac{dy}{dx} = -2x^{-3}$$

b $y = 3\sqrt{x} + \dfrac{2}{x^4} = 3x^{\frac{1}{2}} + 2x^{-4}$

$$\frac{dy}{dx} = 3 \times \tfrac{1}{2}x^{-\frac{1}{2}} + 2 \times (-4)x^{-5}$$

$$= \tfrac{3}{2}x^{-\frac{1}{2}} - 8x^{-5}$$

$$= \frac{3}{2\sqrt{x}} - \frac{8}{x^5}$$

Example 4.5 Differentiate $y = \dfrac{6x^2 + x^3 - 2x}{2x}$ with respect to x.

Step 1: Simplify to obtain the terms in index form.

$$y = \frac{6x^2 + x^3 - 2x}{2x}$$

$$= \frac{6x^2}{2x} + \frac{x^3}{2x} - \frac{2x}{2x}$$

$$= 3x + \tfrac{1}{2}x^2 - 1$$

Step 2: Differentiate term by term.

$$\frac{dy}{dx} = 3 + x$$

Second derivative

The second derivative, $\dfrac{d^2y}{dx^2}$, is obtained by differentiating $\dfrac{dy}{dx}$ with respect to x.

In function notation, if $y = f(x)$, the second derivative is written $f''(x)$.

Example 4.6 Given that $y = 4x^3 + 5x^2 - 3x + 1$, find   **a** $\dfrac{dy}{dx}$ **b** $\dfrac{d^2y}{dx^2}$.

Step 1: Differentiate with respect to x.

$$y = 4x^3 + 5x^2 - 3x + 1$$

a $\dfrac{dy}{dx} = 12x^2 + 10x - 3$

Step 2: Differentiate again.   **b** $\dfrac{d^2y}{dx^2} = 24x + 10$

Example 4.7 If $f(x) = x^3$, find $f''(2)$.

Step 1: Differentiate twice.

$$f(x) = x^3$$
$$f'(x) = 3x^2$$
$$f''(x) = 6x$$

Step 2: Substitute $x = 2$. $\Rightarrow$ $f''(2) = 6 \times 2 = 12$

4.3 Differentiation and gradients

Apply differentiation to gradients.

To find the gradient at a point on a curve, substitute the x-value of the point into the gradient function.

Example 4.8 The sketch shows the curve $y = (x - 3)^2$.

Using differentiation, find the gradient when

a $x = 3$ **b** $x = 4$

Step 1: Form an expanded polynomial.

$$y = (x - 3)^2$$
$$= x^2 - 6x + 9$$

Step 2: Differentiate term by term.

$$\frac{dy}{dx} = 2x - 6$$

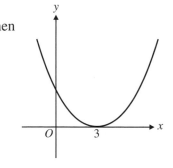

Step 3: Substitute the x-value into the derivative.

a At $x = 3$, $\dfrac{dy}{dx} = 2(3) - 6 = 0$

The gradient when $x = 3$ is 0.

b At $x = 4$, $\dfrac{dy}{dx} = 2(4) - 6 = 2$

The gradient when $x = 4$ is 2.

Tip:
Notice that the tangent at $x = 3$ is horizontal, so you expect the gradient to be zero. Differentiating provides a check.

Tip:
Each term must be in index form before differentiating.

Example 4.9 A curve has equation $y = 2\sqrt{x} - 2x$.

Differentiate y with respect to x and hence find the gradient at the point $(4, -4)$.

Step 1: Write in index form.

$$y = 2\sqrt{x} - 2x = 2x^{\frac{1}{2}} - 2x$$

Step 2: Differentiate term by term.

$$\frac{dy}{dx} = 2 \times \tfrac{1}{2}x^{-\frac{1}{2}} - 2$$
$$= x^{-\frac{1}{2}} - 2$$
$$= \frac{1}{\sqrt{x}} - 2$$

Step 3: Substitute the x-value into the derivative.

When $x = 4$, $\dfrac{dy}{dx} = \dfrac{1}{\sqrt{4}} - 2 = -\tfrac{3}{2}$.

The gradient at $(4, -4)$ is $-\tfrac{3}{2}$.

Example 4.10 It is given that $f(x) = \left(\dfrac{1}{x} - \dfrac{1}{x^2}\right)^2$.

    a Find the derived function, $f'(x)$.

    b Find $f'(1)$ and hence write down the gradient of the curve $y = f(x)$ at the point $(1, 0)$.

Step 1: Write in index form.

a $f(x) = \left(\dfrac{1}{x} - \dfrac{1}{x^2}\right)^2$

$\qquad = (x^{-1} - x^{-2})^2$

$\qquad = x^{-2} - 2x^{-3} + x^{-4}$

> **Recall:**
> $(a - b)^2 = a^2 - 2ab + b^2$

> **Recall:**
> Index laws (Section 1.1).

Step 2: Differentiate term by term.

$f'(x) = -2x^{-3} - 2(-3)x^{-4} + (-4)x^{-5}$

$\qquad = -2x^{-3} + 6x^{-4} - 4x^{-5}$

> **Note:**
> You could write
> $f'(x) = -\dfrac{2}{x^3} + \dfrac{6}{x^4} - \dfrac{4}{x^5}$

Step 3: Substitute the x-value into the derivative.

b $f'(1) = -2(1)^{-3} + 6(1)^{-4} - 4(1)^{-5}$

$\qquad = -2 + 6 - 4$

$\qquad = 0$

The gradient at $(1, 0)$ is given by $f'(1)$.

Hence the gradient at $(1, 0)$ is zero.

SKILLS CHECK **4A: Differentiating and finding gradients**

1 Differentiate with respect to x:

    a $y = x^4$
    b $f(x) = 7x^3 - 2x^2 + 3$
    c $f(x) = (4x + 1)^2$
    d $y = \frac{1}{2}x(2x - 1)$

    e $y = x^{-3}$
    f $y = \dfrac{2}{x^5}$
    g $y = 4x^{\frac{5}{2}}$
    h $y = \dfrac{6}{\sqrt{x}}$

2 Find the derivative, with respect to x, of

    a $y = x(x - 1)^2$
    b $y = \dfrac{3x^4 + 5x}{2}$
    c $y = 2$.

3 It is given that $f(x) = x^2\left(\dfrac{1}{x^4} + \dfrac{2}{x^2}\right)$. Find the derived function, $f'(x)$.

4 It is given that $y = \dfrac{x^2 - x^{\frac{3}{2}}}{2x^2}$.

    a Find $\dfrac{dy}{dx}$.

    b Find the gradient of the curve at $x = 1$.

5 A curve has equation $y = x^2(x - 3)$.

    a Find $\dfrac{dy}{dx}$.

    b Determine the point on the curve where the gradient is -3.

6 Evaluate $f'(4)$, where $f(x) = (2x - 3\sqrt{x})^2$.

7 **a** The variables P and x are connected by the formula $P = (4x - 3)(x + 5)$. Find $\dfrac{dP}{dx}$.

    b Given that $A = 5y - 3y^2$, find $\dfrac{dA}{dy}$.

8 The sketch shows the curve $y = \dfrac{3}{2x^2}$.

The points A, B and C on the curve have x-coordinates -2, -1 and 3 respectively.

Find the gradient of the curve at each of the points A, B and C.

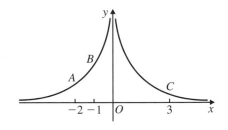

9 Find the values of x for which the gradient of the curve $y = x^2 + 16x^{-2}$ is zero.

10 Find the values of x for which the gradient of the curve $y = \dfrac{3}{x} + 9x$ is the same as the gradient of the line $y = 3x + 1$.

11 Find the gradient of the curve $y = 2\sqrt{x^3} - 4x$ at the origin.

12 Given that $y = \dfrac{x^3 - 5x}{\sqrt{x}}$, show that $\dfrac{dy}{dx} = \dfrac{5(x^2 - 1)}{2\sqrt{x}}$.

13 Given that $y = x^3 + \dfrac{3}{x^3}$, find the value of $\dfrac{d^2y}{dx^2}$ when $x = 1$.

SKILLS CHECK **4A EXTRA** is on the CD.

4.4 Tangents and normals

Apply differentiation to tangents and normals.

The **tangent** at P is the line that touches the curve at P.

The **normal** at P is the line through P perpendicular to the tangent.

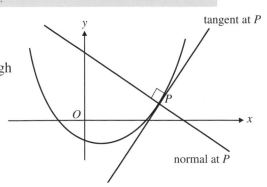

The tangent and normal are perpendicular, so the product of their gradients is -1.

To find the **equation of the tangent** to the curve $y = f(x)$ at the point $P(x_1, y_1)$, find the gradient of the tangent, then substitute into the equation of a line:

$$y - y_1 = m(x - x_1).$$

To find the **equation of the normal**, first find the gradient of the tangent and use it to calculate the gradient of the normal. You can then substitute into the equation of a line, as before.

Note:
If gradient of tangent $= 2$, then gradient of normal $= -\frac{1}{2}$ (Section 3.3).

Recall:
Equation of a straight line (Section 3.2)

Note:
You could use $y = mx + c$.

Example 4.11 A curve has equation $y = x^3 - 6x + 1$. The point P(2, −3) lies on the curve. Giving your answers in the form $ax + by + c = 0$, where a, b and c are integers, find the equation of

a the tangent at P **b** the normal at P

a $y = x^3 - 6x + 1$

Step 1: Find the gradient function.

$$\frac{dy}{dx} = 3x^2 - 6$$

Step 2: Find the gradient of the tangent at P.

When $x = 2$, $\dfrac{dy}{dx} = 3(2)^2 - 6 = 6$

The gradient of the tangent at (2, −3) is 6.

Equation of tangent at P:

Step 3: Use $y - y_1 = m(x - x_1)$.

$$y - (-3) = 6(x - 2)$$
$$y + 3 = 6x - 12$$
$$6x - y - 15 = 0$$

> **Note:**
> $(x_1, y_1) = (2, -3)$ and $m = 6$.

Step 1: Find the gradient the of normal.

b Since product of gradients $= -1$, gradient of normal $= -\frac{1}{6}$.

Equation of normal:

Step 2: Use $y - y_1 = m(x - x_1)$.

$$y - (-3) = -\tfrac{1}{6}(x - 2)$$
$$y + 3 = -\tfrac{1}{6}(x - 2)$$
$$6(y + 3) = -x + 2$$
$$x + 6y + 16 = 0$$

> **Note:**
> $(x_1, y_1) = (2, -3)$ and $m = -\frac{1}{6}$.

Example 4.12 A curve has equation $y = \dfrac{4}{x^3} - \dfrac{x^2}{4}$ and the point P$(2, -\frac{1}{2})$ lies on the curve.

a The equation of the tangent at P is $px + qy = r$, where p, q and r are integers. Find the values of p, q and r.

b Find an equation of the normal at P.

> **Tip:**
> Write all terms in index form before differentiating.

Step 1: Differentiate to find the gradient function.

a $y = \dfrac{4}{x^3} - \dfrac{x^2}{4} = 4x^{-3} - \tfrac{1}{4}x^2$

$$\frac{dy}{dx} = -12x^{-4} - \tfrac{1}{2}x$$

> **Tip:**
> Take care with negatives.

Step 2: Substitute the x-value to get the gradient of the tangent at P.

When $x = 2$, $\dfrac{dy}{dx} = (-12 \times 2^{-4}) - (\tfrac{1}{2} \times 2) = -\tfrac{7}{4}$.

The gradient of the tangent at P is $-\tfrac{7}{4}$.

Equation of the tangent at P:

Step 3: Use $y - y_1 = m(x - x_1)$.

$$y - (-\tfrac{1}{2}) = -\tfrac{7}{4}(x - 2)$$
$$y + \tfrac{1}{2} = -\tfrac{7}{4}(x - 2)$$

Step 4: Rearrange to the required format.

$$4(y + \tfrac{1}{2}) = -7(x - 2)$$
$$4y + 2 = -7x + 14$$
$$7x + 4y = 12$$

So $p = 7$, $q = 4$, $r = 12$.

> **Tip:**
> Rearrange the equation to the given format.

Step 1: Find the gradient of the normal at P.

b At P, gradient of tangent $= -\tfrac{7}{4}$

$\Rightarrow$ gradient of normal $= \tfrac{4}{7}$

> **Tip:**
> $m_1 \times m_2 = -1$, so find the negative reciprocal.

Step 2: Use $y - y_1 = m(x - x_1)$.

Equation of normal at P:

$$y - (-\tfrac{1}{2}) = \tfrac{4}{7}(x - 2)$$
$$y + \tfrac{1}{2} = \tfrac{4}{7}(x - 2)$$

> **Tip:**
> The equation may be left in this format if one is not specified.

4.5 Increasing and decreasing functions

Apply differentiation to increasing and decreasing functions.

Consider a function f(x).

If, as the x-value increases, the corresponding value of f(x) increases, the function is an **increasing function**.

The gradient of the curve $y = $ f(x) is positive.

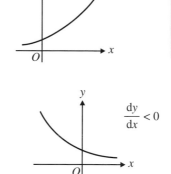

If, as the x-value increases, the corresponding value of f(x) decreases, the function is a **decreasing function**.

The gradient of the curve $y = $ f(x) is negative.

Note:
Curves can be both increasing and decreasing, depending on which x-values you are considering.

Example 4.13 Find the set of values of x for which the curve $y = (x - 3)^2$ is increasing.

Step 1: Find $\dfrac{dy}{dx}$.
$$y = (x - 3)^2$$
$$= x^2 - 6x + 9$$
$$\frac{dy}{dx} = 2x - 6$$

Step 2: Set $\dfrac{dy}{dx} > 0$. For an increasing function, $\dfrac{dy}{dx} > 0$

Step 3: Solve the inequality.
$$\Rightarrow 2x - 6 > 0$$
$$x > 3$$

Recall:
Inequalities (Section 2.8).

Note:
This agrees with the sketch in Example 4.8.

Example 4.14 Find the set of values for which the function f(x) = $4x^2 - x$ is decreasing.

Step 1: Find f'(x). f(x) = $4x^2 - x \Rightarrow$ f'(x) = $8x - 1$

Step 2: Set f'(x) < 0. For a decreasing function, f'(x) < 0
$$\Rightarrow 8x - 1 < 0$$

Step 3: Solve the inequality.
$$x < \tfrac{1}{8}$$

SKILLS CHECK **4B: Applications of differentiation**

1 Find the equation of the tangent to the curve $y = x^3 - 3$ at the point $(1, -2)$.

2 A curve has equation $y = x^3 + 2x - 1$.

   a The curve goes through the point $P(1, q)$. Find q.

   b Find the gradient at P.

   c Find the equation of the tangent at P, giving your answer in the form $y = mx + c$.

   d Find the equation of the normal at P, giving your answer in the form $ax + by + c = 0$.

3 Find the set of values of x for which the function is **i** increasing, **ii** decreasing:

   a f(x) = $(x - 2)^2$ **b** f(x) = $x^2 + 8$ **c** g(x) = $x^3 - 3x + 2$

4 The sketch shows the curve $y = x^2 - 2x + 3$ and the normal to the curve at $(0, 3)$.

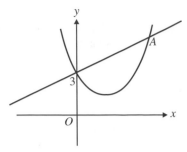

The normal intersects the curve again at A.

  a Find the gradient of the tangent at $(0, 3)$.

  b Find an equation of the normal at $(0, 3)$.

  c Find the x-coordinate of A.

5 **a** Find an equation of the normal to the curve $y = x^2 + 3$ at the point where $x = -1$.

  b By solving simultaneous equations, find the coordinates of the point where the normal meets the curve again.

  c Draw a sketch to illustrate the information.

6 A curve has equation $y = (x - 2)^2 - 3$.

  a Find an equation of the tangent at the point on the curve where $x = -1$.

  b Find the coordinates of the point on the curve where the normal is parallel to the line $y = \frac{1}{2}x + 5$.

7 It is given that $f(x) = (2x - 1)(3x + 1)$. Find the values of x for which $f(x)$ is an increasing function.

8 A curve has equation $y = x^3 - 8x + 4$. The point P has x-coordinate 2.

  a Find an equation of the tangent to the curve at P.

  b The tangent crosses the x-axis at A and the y-axis at B. Find the coordinates of A and B.

  c Find the area of triangle OAB, where O is the origin.

9 A curve has equation $y = 2x^{\frac{3}{2}} - 4x^{\frac{5}{2}} + 2x$.

  a Find $\dfrac{dy}{dx}$.

  b Show that the equation of the tangent to the curve at $(1, 0)$ is $y + 5x = 5$.

  c Find the equation of the normal at the point $(1, 0)$, writing your answer in the form $ax + by + c = 0$, where a, b and c are integers.

10 Find the equation of the normal to the curve $y = \dfrac{4}{x} + x^2$ at the point where $x = 1$, giving your answer in the form $ax + by + c = 0$, where a, b and c are integers.

11 Find the equation of the tangent to the curve $y = 40\sqrt{x}$ at the point where the gradient of the curve is 5, writing your answer in the form $y = mx + c$.

12 It is given that $f(x) = \dfrac{1}{\sqrt[3]{x}}$.

  a Show that $f'(8) = -\frac{1}{48}$.

  b Hence find the equation of the tangent to the curve $y = f(x)$ at the point where $x = 8$, giving your answer in the form $ax + by + c = 0$, where a, b and c are integers.

SKILLS CHECK **4B EXTRA** is on the CD

Application of differentiation to the location of stationary points.

At a **stationary point** on a curve, the gradient is zero, so $\frac{dy}{dx} = 0$.

In Core 1, the stationary points studied are either maximum turning points or minimum turning points.

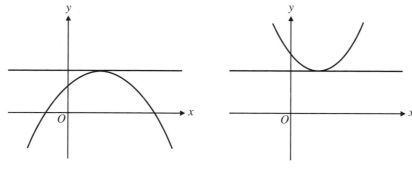

Note:
The tangent at a stationary point is parallel to the x-axis.

Maximum point
(the turn is at the top)

Minimum point
(the turn is at the bottom)

Example 4.15 By differentiating, find the coordinates of the stationary point of the curve $y = (x - 2)^2 + 3$.

Recall:
Expand and simplify first.

Step 1: Find $\frac{dy}{dx}$.
$$y = (x - 2)^2 + 3$$
$$= x^2 - 4x + 7$$
$$\frac{dy}{dx} = 2x - 4$$

Step 2: Put $\frac{dy}{dx} = 0$ and solve for x.
$$\frac{dy}{dx} = 0 \text{ when } 2x - 4 = 0$$
So $x = 2$.

Step 3: Substitute the x-value into equation of curve to find the y-coordinate.
When $x = 2$, $y = (2 - 2)^2 + 3 = 3$
So there is a stationary point at $(2, 3)$.

Note: This confirms what you would know from work on quadratic functions and completing the square. For the curve $y = (x - p)^2 + q$, the minimum value is q and it occurs when $x = p$. This indicates that there is a minimum turning point at (p, q).

Recall:
Completing the square (Section 2.5) and sketching curves (Section 3.7).

Determining the nature of a stationary point

Here are two methods for determining the nature of a stationary point.

Method 1

Investigate the value of $\frac{dy}{dx}$ for x-values immediately to the left and to the right of the point.

If, as x increases,

- $\frac{dy}{dx}$ goes from positive to zero to negative, there is a maximum point

- $\frac{dy}{dx}$ goes from negative to zero to positive, there is a minimum point.

Note:
Another method involves investigating the y-value immediately to the left and right of the stationary point.

Note:
Direction of slope near stationary point:

Maximum:

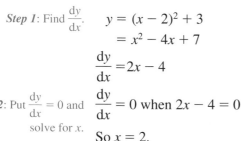

Minimum:

Method 2

Consider the sign of $\dfrac{d^2y}{dx^2}$ at the stationary point:

Tip:
Method 2 is often a quicker approach.

- $\dfrac{dy}{dx} = 0$ and $\dfrac{d^2y}{dx^2} > 0 \Rightarrow$ minimum turning point

- $\dfrac{dy}{dx} = 0$ and $\dfrac{d^2y}{dx^2} < 0 \Rightarrow$ maximum turning point.

Note that if $\dfrac{dy}{dx} = 0$ and $\dfrac{d^2y}{dx^2} = 0$, Method 2 fails and Method 1 is advised.

Example 4.16 **a** Find the coordinates of the stationary points on the curve $y = x^3 - 3x + 2$.

b Investigate their nature.

Tip:
Notice that the question suggests there is more than one.

Step 1: Find $\dfrac{dy}{dx}$.

a $y = x^3 - 3x + 2 \Rightarrow \dfrac{dy}{dx} = 3x^2 - 3$

Step 2: Put $\dfrac{dy}{dx} = 0$ and solve for x.

$\dfrac{dy}{dx} = 0$ when $3x^2 - 3 = 0$

$$3(x - 1)(x + 1) = 0$$
$$\Rightarrow \quad x = 1 \text{ or } x = -1$$

Step 3: Substitute x-value into the equation of the curve.

When $x = 1$, $y = 1^3 - 3(1) + 2 = 0$

When $x = -1$, $y = (-1)^3 - 3(-1) + 2 = 4$

Tip:
Take care with negatives.

Step 4: State the coordinates.

There are stationary points at $(1, 0)$ and $(-1, 4)$.

Step 5: Find $\dfrac{d^2y}{dx^2}$.

b Using the second derivative:

$$\dfrac{dy}{dx} = 3x^2 - 3 \Rightarrow \dfrac{d^2y}{dx^2} = 6x$$

Step 6: Substitute x-value to find nature.

When $x = 1$, $\dfrac{d^2y}{dx^2} = 6(1) = 6 > 0 \Rightarrow$ minimum point.

When $x = -1$, $\dfrac{d^2y}{dx^2} = 6(-1) = -6 < 0 \Rightarrow$ maximum point.

There is a minimum turning point at $(1, 0)$ and a maximum turning point at $(-1, 4)$.

Tip:
Summarise the information, so that it is easier for the examiner to see.

Example 4.17 Find the coordinates of the stationary point of the curve $y = x^4 + 2$ and determine its nature.

Step 1: Find $\dfrac{dy}{dx}$.

$y = x^4 + 2 \Rightarrow \dfrac{dy}{dx} = 4x^3$

Step 2: Put $\dfrac{dy}{dx} = 0$ and solve for x.

$\dfrac{dy}{dx} = 0$ when $4x^3 = 0$, so $x = 0$

When $x = 0$, $y = 0^4 + 2 = 2$

Step 3: Substitute x-value into the equation of the curve.

There is a stationary point at $(0, 2)$.

Step 4: Find second derivative.

Using the second derivative:

$$\dfrac{d^2y}{dx^2} = 12x^2$$

When $x = 0$, $\dfrac{d^2y}{dx^2} = 12(0) = 0$

The second derivative method has broken down, so use Method 1 near $x = 0$.

Step 5: Calculate the value of $\dfrac{dy}{dx}$ close to the stationary point.

When $x = -1$, $\dfrac{dy}{dx} = 4x^3 = 4(-1)^3 = -4$

When $x = 1$, $\dfrac{dy}{dx} = 4(1)^3 = 4$

Tip:
You must evaluate the gradient to show that you have done the calculation.

Step 6: Draw a line representing the slope of the gradient.

	$x = -1$	$x = 0$	$x = 1$
Sign of $\dfrac{dy}{dx}$	−	0	+
Direction of gradient	╲	──	╱

Tip:
Do not use x-values too far away from the stationary point as you may pass another stationary point!

Step 7: State the nature of the stationary point.

There is a minimum turning point at $(0, 2)$.

Example 4.18 A curve has equation $y = \dfrac{1}{x} + 32x^2$. Find the coordinates of the stationary point on the curve and determine its nature.

Step 1: Write in index form.

$y = \dfrac{1}{x} + 32x^2 = x^{-1} + 32x^2$

Tip:
Write in index form before differentiating.

Step 2: Find $\dfrac{dy}{dx}$.

$\dfrac{dy}{dx} = -x^{-2} + 64x = -\dfrac{1}{x^2} + 64x$

Step 3: Set $\dfrac{dy}{dx} = 0$ and solve for x.

$\dfrac{dy}{dx} = 0$ when $-\dfrac{1}{x^2} + 64x = 0$

$$64x = \dfrac{1}{x^2}$$

$(\times x^2)$

$$x^3 = \tfrac{1}{64}$$

$$x = \sqrt[3]{\tfrac{1}{64}} = \tfrac{1}{4}$$

Step 4: Calculate the y-coordinate.

When $x = \tfrac{1}{4}$, $y = \dfrac{1}{x} + 32x^2 = \dfrac{1}{\tfrac{1}{4}} + 32 \times \left(\tfrac{1}{4}\right)^2 = 6$.

Hence there is a stationary point at $\left(\tfrac{1}{4}, 6\right)$.

Step 5: Find $\dfrac{d^2y}{dx^2}$ and substitute the x-value.

$\dfrac{dy}{dx} = -x^{-2} + 64x \Rightarrow \dfrac{d^2y}{dx^2} = 2x^{-3} + 64$

When $x = \tfrac{1}{4}$, $\dfrac{d^2y}{dx^2} = 2 \times \left(\tfrac{1}{4}\right)^{-3} + 64 = 192$.

Since $\dfrac{d^2y}{dx^2} > 0$, $\left(\tfrac{1}{4}, 6\right)$ is a minimum turning point.

Tip:
You must include sufficient working to show whether the second differential is positive or negative.

Example 4.19 A cylindrical tin, closed at both ends, is made from thin sheet metal. The radius of the base of the cylinder is r cm and the volume of the tin is 1024π cm^3.

a Show that the total surface area, S cm^2, of the cylinder is given by

$$S = \dfrac{2048\pi}{r} + 2\pi r^2.$$

b Find the value of r that gives a minimum total surface area and state the value of this surface area in terms of π.

Step 1: Draw a diagram showing given information.	**a** Let the height be h cm.	

a Let the height be h cm.

Volume $= 1024\pi$

Step 2: Write unknown measures in terms of the given variable.

$\Rightarrow \pi r^2 h = 1024\pi$

$$h = \frac{1024}{r^2}$$

Tip:
Use the condition that the volume is 1024π to express h in terms of r.

Step 3: Form an expression in r for the surface area.

$S = 2\pi r h + 2\pi r^2$

$$= 2\pi r \left(\frac{1024}{r^2}\right) + 2\pi r^2$$

$$= \frac{2048\pi}{r} + 2\pi r^2$$

Tip:
Find the curved surface area and the area of the two circular ends.

Step 4: Find $\dfrac{dS}{dr}$.

b $S = 2048\pi r^{-1} + 2\pi r^2$

$$\frac{dS}{dr} = -2048\pi r^{-2} + 4\pi r$$

Tip:
Write all terms in index form before differentiating.

Step 5: Set $\dfrac{dS}{dr} = 0$ and solve for r.

$\dfrac{dS}{dr} = 0$ when $-2048\pi r^{-2} + 4\pi r = 0$

$$4\pi r = \frac{2048\pi}{r^2}$$

$(\times r^2)$ $\qquad\qquad\qquad 4\pi r^3 = 2048\pi$

$(\div 4\pi)$ $\qquad\qquad\qquad r^3 = \dfrac{2048\pi}{4\pi} = 512$

$$r = \sqrt[3]{512} = 8$$

Tip:
$512 = 2^9$ so $\sqrt[3]{512}$
$= \sqrt[3]{2^9} = 2^3 = 8$.

There is a stationary value when $r = 8$.

Step 6: Check the nature of the stationary value.

$$\frac{dS}{dr} = -2048\pi r^{-2} + 4\pi r \Rightarrow \frac{d^2S}{dr^2} = 4096\pi r^{-3} + 4\pi$$

When $r = 8$, $\dfrac{d^2S}{dr^2} = 4096\pi \times 8^{-3} + 4\pi = 37.6\ldots > 0$

Tip:
You will not be allowed to use a calculator in C1, but, since all the terms are positive, it is obvious that $\dfrac{d^2S}{dr^2} > 0$.

So S has a minimum value when $r = 8$.

Step 7: Substitute for r into S.

When $r = 8$, $S = \dfrac{2048\pi}{8} + 2\pi \times 8^2 = 384\pi$.

The minimum surface area is 384π cm^2.

Tip:
Remember to give your answer in terms of π and include the units.

SKILLS CHECK **4C: Stationary points and problems**

1 Find the coordinates of the stationary points of the given curves and determine their nature.

    a $y = x^2 + x^3$ **b** $y = x^3 - 3x$ **c** $y = 2x^3 + 3x^2 - 12x + 6$

2 Find the x-coordinates of the stationary points on the curve $y = (1 - x^2)(1 - 4x)$ and determine their nature.

3 A builder wishes to make a rectangular enclosure around a garden. The house is to form one of the sides; this side has length $4x$ metres. The other three sides are to be fenced with a total of 2000 m of fencing.

Show that the area of the garden, A, is given by $A = 4000x - 8x^2$.

Hence, find the maximum area of the garden, verifying that the value you have found is a maximum.

4 The sum of two variable positive numbers is 200.

Let x be one of the numbers, and let the product of these two numbers be y.

Find the maximum value of y.

5 The diagram shows a square piece of card, with sides of length 6 cm. A smaller square, of side x cm, is cut from each corner as shown. The card is folded along the dotted lines to make an open box.

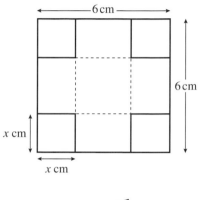

a The volume of the box is V cm^3. Show that
$V = 36x - 24x^2 + 4x^3$.

b Given that x can vary, find the values of x for which $\dfrac{dV}{dx} = 0$.

c Show that one of the values of x gives a maximum value for V and find the maximum value of V.

6 The diagram shows a triangular prism whose cross-section is a right-angled triangle with sides $3x$ cm, $4x$ cm and $5x$ cm. The length of the prism is l cm. The sum of the edges of the prism is 90 cm and it has a volume of V cm^3.

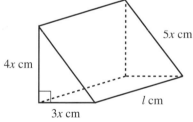

a Show that $l = 30 - 8x$.

b Show that $V = 12x^2(15 - 4x)$.

c Find $\dfrac{dV}{dx}$ and hence calculate the value of x that gives a maximum volume.

d The surface area of the prism is A cm^2. Express A in terms of x.

e Show that the surface area is greatest when $x = 2\frac{1}{7}$ cm.

7 A curve has equation $y = \dfrac{x^2 - 2x^3}{x^5}$.

Find the x-coordinate of the stationary point on the curve and determine the nature of the stationary point.

8 The diagram shows a sketch of the curve $y = 12x^{\frac{1}{2}} - 2x^{\frac{3}{2}}$.

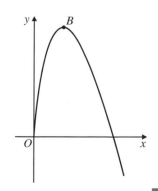

a Find the x-coordinate of B.

b Hence state the maximum value of y, leaving your answer in the form $a\sqrt{2}$, where a is an integer to be found.

9 A closed cuboid is to be made from thin cardboard. The base of the cuboid is a rectangle with width x cm. The length of the base is twice the width and the volume of the cuboid is 1944 cm^3. The surface area of the cuboid is S cm^2.

a Show that $S = 4x^2 + 5832x^{-1}$.

b Given that x can vary, show that the value of x that makes the surface area a minimum is 9.

c Find the minimum value of the surface area.

SKILLS CHECK **4C EXTRA** is on the CD

Examination practice   Differentiation

1

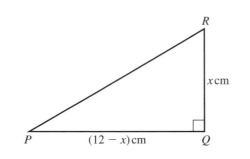

The diagram shows the triangle PQR with $PQ = (12 - x)$ cm, $QR = x$ cm and angle $PQR = 90°$. The area of the triangle is A cm^2.

i Show that $A = 6x - \frac{1}{2}x^2$.

ii Find $\dfrac{dA}{dx}$.

iii Hence find the greatest possible area of triangle PQR, showing that it is the greatest and not the least.   [OCR Nov 2003]

2 It is given that $f(x) = x^3 - 4x^2 - 3x + 1$.

i Find $f'(x)$.

ii Find $f''(x)$.

iii Find the x-coordinates of the stationary points on the curve $y = x^3 - 4x^2 - 3x + 1$.

iv Determine whether each stationary point is a maximum or a minimum point.   [OCR Jan 2004]

3 i Find the equation of the tangent to the curve $y = x^2 + 4x + 3$ at the point whose x-coordinate is 1. Give your answer in the form $y = mx + c$.

The line $y = 4x + k$ is a tangent to the curve $y = x^2 + 4x + 3$ at a point P.

ii Find the value of k.

iii Find an equation of the normal to the curve $y = x^2 + 4x + 3$ at the point P.   [OCR Jan 2004]

4 i Sketch the graph of $y = \dfrac{1}{x}$, where $x \neq 0$, showing the parts of the graph corresponding to both positive and negative values of x.

ii Describe fully the geometrical transformation that transforms the curve $y = \dfrac{1}{x}$ to the curve $y = \dfrac{1}{x+2}$. Hence sketch the curve $y = \dfrac{1}{x+2}$.

iii Differentiate $\dfrac{1}{x}$ with respect to x.

iv Use parts **ii** and **iii** to find the gradient of the curve $y = \dfrac{1}{x+2}$ at the point where it crosses the y-axis.

[OCR Spec paper]

5 Differentiate $3x^5 + \dfrac{1}{2x^2}$ with respect to x.

6 Given that $f(x) = 2\sqrt{x} - \dfrac{1}{x}$, find $f'(\tfrac{1}{4})$.

7 Find $\dfrac{dy}{dx}$ in each of the following cases:

i $y = 4x^3 - 1$,

ii $y = x^2(x^2 + 2)$,

iii $y = \sqrt{x}$

[OCR Spec paper]

8 i Find the coordinates of the stationary points on the curve $y = 2x^3 - 3x^2 - 12x - 7$.

ii Determine whether each stationary point is a maximum point or a minimum point.

iii By expanding the right-hand side, show that
$$2x^3 - 3x^2 - 12x - 7 = (x + 1)^2(2x - 7).$$

iv Sketch the curve $y = 2x^3 - 3x^2 - 12x - 7$, marking the coordinates of the stationary points where the curve meets the axes.

[OCR Spec paper]

9 Given that $f(x) = 3x^2 - 5x - 2$, find $f'(x)$.

[OCR May 2003]

Practice exam paper

Answer **all** questions.
Time allowed: 1 hour 30 minutes
A calculator is **not** to be used in this paper.

1 The formula for the volume of a sphere is $V = \frac{4}{3}\pi r^3$.

Find $\dfrac{dV}{dr}$. *(2 marks)*

2 Find, in its simplest form, the product of $a^{\frac{1}{3}} - b^{\frac{1}{3}}$ and $a^{\frac{2}{3}} + a^{\frac{1}{3}}b^{\frac{1}{3}} + b^{\frac{2}{3}}$. *(3 marks)*

3 Solve the equation $x^6 - 7x^3 - 8 = 0$. *(4 marks)*

4 i Express $\dfrac{1}{2 + \sqrt{3}}$ in its simplest surd form. *(2 marks)*

  ii Given that the approximate value of $\sqrt{3}$ is 1.7321 correct to five significant figures,

    calculate the approximate value of $\dfrac{1}{\sqrt{12}}$ correct to three significant figures. *(3 marks)*

5 The point A has coordinates $(-4, -5)$ and the point B has coordinates $(2, 7)$.
The point M is the midpoint of AB.

  i Find the coordinates of M. *(2 marks)*

  ii Calculate the distance AB, giving your answer in its simplest form. *(3 marks)*

  iii Find the equation of the straight line through M perpendicular to AB. *(3 marks)*

6 A circle has equation

$$x^2 + y^2 + 4x + 2y - 4 = 0.$$

  i By completing the square, find the coordinates of the centre of the circle and
its radius. *(5 marks)*

  ii Calculate in surd form the distance of the centre of the circle from the origin.
State whether the origin lies inside or outside the circle. *(3 marks)*

7 The parabola $y = x^2 + k$ and the line $y = kx - 3$ do **not** intersect. Find the range of
possible values of k. *(6 marks)*

  On a single diagram, sketch the parabola and the line for the case when $k = -1$. *(3 marks)*

8 i A curve has the equation $y = 3\sqrt{x}$, for $x \geq 0$. Sketch the curve. *(1 mark)*

  ii Show that the point $(4, 6)$ lies on the curve. *(1 mark)*

  iii Differentiate $3\sqrt{x}$, and hence show that the gradient of the curve at $(4, 6)$ is $\frac{3}{4}$. *(2 marks)*

  iv Find the equation of the tangent to the curve at $(4, 6)$. *(2 marks)*

  v A straight line, parallel to the tangent at $(4, 6)$, is drawn through the origin.
State the equation of this line, and find the coordinates of the points where
the line and the curve intersect. *(4 marks)*

9 i Find the value of $\dfrac{dy}{dx}$ when $y = \sqrt[3]{x^2}$ and $x = 64$. *(3 marks)*

ii Find $f'(x)$ when $f(x) = (x-1)(x-3)$. *(3 marks)*

iii Find $\dfrac{d^2x}{dt^2}$ when $x = \dfrac{4t^2 - 3}{t}$. *(4 marks)*

10 The function $f(x) = x^3 + 6x^2 - 15x$.

i Find $f'(x)$ and $f''(x)$. *(3 marks)*

ii Find the coordinates of the turning points of the curve $y = f(x)$, and evaluate $f''(x)$ at these points. *(5 marks)*

iii The x-axis is a tangent to the curve $y = f(x) + a$. Find the two possible values of a. In two separate diagrams, sketch the corresponding graphs. *(5 marks)*

Answers

SKILLS CHECK 1A (page 2)

1 a $6x^4y^4$ b $16a^8$ c $7p^{-1}q^2$

2 a 2 b $\frac{1}{9}$ c $\frac{1}{3}$ d 64 e 2

3 $12a^5b^{-\frac{3}{2}}$

4 a x^2 b x^{-1} c x^5

5 $p^{\frac{1}{3}}$

6 a 2^{2x} b 2^{3x-3}

7 a 4 b 27

8 a 13 b 0

SKILLS CHECK 1B (page 5)

1 a $5\sqrt{2}$ b $4\sqrt{2}$ c $7\sqrt{2}$ d $7\sqrt{3}$ e $7\sqrt{7}-4\sqrt{5}$

2 a $11\sqrt{3}$ b $2\sqrt{2}$ c $5-2\sqrt{6}$ d 50 e $2\sqrt{2}$

   f 3 g $\frac{\sqrt{3}}{3}$ h $\sqrt{6}-2$

3 a $12+2\sqrt{11}$ b 10 c $\frac{6}{5}+\frac{\sqrt{11}}{5}$

4 $p=\frac{3}{2}$

5 $4\sqrt{3}, 2\sqrt{3}$ b $6\sqrt{3}$

6 a $8+2\sqrt{7}$ b $\frac{4}{3}+\frac{1}{3}\sqrt{7}$

7 a $p=7, q=1$ b $r=\frac{7}{11}, s=\frac{1}{11}$

Exam practice 1 (page 6)

1 i 4 ii $\frac{1}{5}$

2 i $\frac{1}{16}$ ii 8 iii 6

3 a $23\sqrt{6}$ b -1

4 $2+5\sqrt{5}$

5 $\frac{2}{3}$

6 8

7 $\frac{22}{23}-\frac{9}{23}\sqrt{2}$

8 $\frac{x^2}{8}$

9 $47+6\sqrt{10}$

10 a x^{-1} b x^4 c $x^{\frac{3}{4}}$

11 $16x^2$

12 a i $4\sqrt{3}$ ii $2\sqrt{3}$ b $2\sqrt{3}$

13 a $a=13, b=-2$ b $p=\frac{13}{7}, q=-\frac{2}{7}$

14 $\frac{1}{3}$

SKILLS CHECK 2A (page 9)

1 a $x^3+12x^2+23x-2$ b $x^3+3x^2-11x-3$
   c $-3x^3+2x^2+23x$

2 a $-x^2-18x+5$ b $4x^3+14x^2-7x+4$
   c $x^3-5x^2-9x+45$ d $2x^3+3x^2-23x-12$

3 a $a=2, b=5$ b $a=3, b=-1, c=5$
   c $a=-4, b=-\frac{3}{2}, c=-2$ or $a=3, b=2, c=-2$

SKILLS CHECK 2B (page 13)

1 a $x(x+5)$ b $(x-1)(x-1)$ c $(a+4)(a-4)$
   d $(x+1)(x-6)$ e $(x+15)(x-2)$ f $2x(x-4)$

2 a $(2x+3)(x+2)$ b $(5x+1)(x-3)$ c $x(x+3)(x-7)$
   d $(3y-2)(y+2)$ e $4(1-2x)(5x+3)$ f $(2x+5)(2x-5)$

3 a $(x+3)^2-1$ b $(x-6)^2-39$ c $\left(x+\frac{5}{2}\right)^2-\frac{33}{4}$

4 a $2\left(x+\frac{7}{4}\right)^2-\frac{1}{8}$ b $5\left(x-\frac{7}{5}\right)^2-\frac{64}{5}$ c $x(x+3)(x-7)$

   d $3\left(y+\frac{2}{3}\right)^2-\frac{16}{3}$ e $-40\left(x+\frac{1}{20}\right)^2+\frac{121}{10}$ f $4x^2-25$

5 a $p=2, q=3$ b $(2,3)$ c Minimum   d $x=2$

6 a $4(x+1)^2-3$ b $(-1,-3)$, minimum   c $x=-1$

7 a $18-(x+2)^2$ b 18 c $(-2, 18)$

SKILLS CHECK 2C (page 16)

1 a $x=2, x=-3$ b $x=\frac{1}{4}, x=-\frac{3}{2}$ c $x=0, x=-5$

2 a $x=-5, x=-1$ b $x=8, x=3$ c $x=0, x=6$
   d $x=-1, x=6$ e $x=-2, x=3$ f $x=\pm 6$

3 a $\left(x+\frac{3}{2}\right)^2-\frac{29}{4}$ b $x=\frac{-3\pm\sqrt{29}}{2}$

4 a $2\left(x-\frac{3}{4}\right)^2-\frac{25}{8}$ b $x=-\frac{1}{2}, x=2$

5 $x=\frac{-3\pm\sqrt{69}}{10}$

6 a $x=-1, x=4$ b $x=\frac{7\pm\sqrt{13}}{4}$

7 $x=-\frac{1}{3}\pm\frac{1}{3}\sqrt{13}$

8 a 2 b 2 c 1 (repeated)   d 2

9 -56

SKILLS CHECK 2D (page 18)

1 a $x=4, y=1$ b $x=3, y=2$ c $a=3, b=-2$

2 a $x=-6, y=4$ b $a=11, b=3$ c $p=-9, q=3$

3 b $x=4, y=-\frac{2}{3}$

4 a $x=2, y=8$ or $x=-6, y=16$
   b $x=\frac{2}{3}, y=2$ or $x=-1, y=-3$
   c $x=0, y=1$ or $x=-1, y=0$

5 width $=3$ cm, length $=6$ cm

6 b $2a+b=4$ c $y=3x^2-2x$

SKILLS CHECK 2E (page 21)

1 a $x>3$ b $x>-1$ c $x\geqslant 8\frac{2}{3}$ d $y>1$
   e $x>6$ f $x\leqslant 14.5$

2 a $3<x<9$ b $-6\leqslant x\leqslant 10$

3 $a=3$

4 a $y<-2, y>2$ b $-7\leqslant x\leqslant 7$ c $x\leqslant -\sqrt{5}, x\geqslant\sqrt{5}$
   d $-3<x<3$ e $x<-1, x>3$
   f $-2-\sqrt{5}\leqslant x\leqslant -2+\sqrt{5}$

5 a $p=2, q=-9$ b $x\leqslant -5, x\geqslant 1$

6 a $-4<x<3$ b $x\leqslant -2\frac{1}{2}, x\geqslant\frac{2}{3}$ c $x\leqslant -5, x\geqslant 4$
   d $-5<p<-2$ e $x\leqslant -2, x\geqslant\frac{3}{2}$

7 a $p=2, q=10$ c $2-\sqrt{10}<x<2+\sqrt{10}$

8 a $-6<k<6$

9 a k^2-16 b $k<-4, k>4$

SKILLS CHECK 2F (page 23)

1 a 4 b 1 or 2

2 a ± 2 or ± 3 b 125 c $\frac{16}{9}$ d $\frac{8}{27}$ or 1

Exam practice 2 (page 23)

1 i $a=4, b=5, c=-3$ ii $x=-5$

2 i k^2-4k ii $0<k<4$

3 i -12
   ii 0, parabola $\cap$ shaped, so does not meet or cross the x-axis

4 $x=\frac{1}{4}, 9$

5 i $a = 3, b = 11$ **ii** $11, x = -3$ **iii** $\frac{1}{11}$

6 i $3(x + \frac{2}{3})^2 - \frac{1}{3}$ **ii** $(-\frac{2}{3}, -\frac{1}{3})$

7 ii $x = 4, 25$

8 $0 < x < \frac{4}{3}$

9 $x > \frac{5}{3}$

10 $x = -2, y = 10$

11 $2x^3 - 5x^2 - 37x + 60$

12 $x = -\frac{1}{3}, y = \frac{7}{3}$ or $x = 3, y = -1$

13 $x = \pm 2$

14 $2(x + 3)^2 - 5$

SKILLS CHECK 3A (page 28)

1 a $3x - y - 2 = 0$ **b** $3x + 2y - 6 = 0$
  c $2x + 3y - 20 = 0$ **d** $6x + 5y - 20 = 0$

2 a $y = \frac{4}{5}x - \frac{8}{5}$ **b** Gradient $= -\frac{2}{3}$, y-intercept $= 2$

3 a i -1 **ii** $(\frac{1}{2}, \frac{5}{2})$ **iii** $7\sqrt{2}$
  b i $\frac{5}{3}$ **ii** $(1, 1)$ **iii** $2\sqrt{34}$
  c i $\frac{1}{3}$ **ii** $(\frac{1}{2}, -\frac{3}{2})$ **iii** $\sqrt{10}$

4 a $y = \frac{3}{5}x + \frac{2}{5}$ **b** $y = -3x + 12$ **c** $y = -x + 3$

6 b $(-3, -1)$ **c** 15 units^2

SKILLS CHECK 3B (page 30)

1 a Perpendicular **b** Parallel **c** Neither

3 $y = \frac{2}{3}x + 3$

4 $y = -\frac{5}{4}x - 2$

5 $y = 5x - 16$

6 $y = 3x - 7$

7 $5x + 3y + 19 = 0$

8 $x + 5y - 12 = 0$

9 a $\frac{5}{2}$ **b** $y = \frac{5}{2}x - \frac{29}{2}$ **c** $-\frac{1}{2}$ **d** $y = -\frac{x}{2} + \frac{1}{2}$ **e** $(5, -2)$

SKILLS CHECK 3C (page 32)

1 a $(x - 3)^2 + (y + 2)^2 = 16$ **b** $(x + 5)^2 + y^2 = 25$

2 a $(2, 4), 6$ **b** $(0, -3), 4$

3 a $x^2 + y^2 + 2x - 4y - 20 = 0$ **b** $x^2 + y^2 + 6x - 8y = 0$
  c $x^2 + y^2 - 4x - 12y + 20 = 0$ **d** $x^2 + y^2 + 4x - 4y - 8 = 0$

4 a $(x - 1)^2 + (y - 2)^2 = 5^2, (1, 2), 5$
  b $(x + 5)^2 + (y + 12)^2 = 13^2, (-5, -12), 13$
  c $(x - 3)^2 + (y + 5)^2 = 4^2, (3, -5), 4$
  d $(x + 1)^2 + (y - 3)^2 = 7, (-1, 3), \sqrt{7}$

5 $(x - 2)^2 + (y + 5)^2 = 5^2$

6 $(x - 1)^2 + (y + 4)^2 = 10$

7 $(1, 1), \sqrt{5}; (7, 4), 2\sqrt{5}$

8 a $-\frac{2}{3}$ **b** $y = \frac{3}{2}x - \frac{11}{2}$

9 a $(x - 2)^2 + (y + 3)^2 = 4^2$ **b** $4, (2, -3)$ **c** $\begin{pmatrix} 2 \\ -3 \end{pmatrix}$

10 $x^2 + y^2 - 10x + 6y - 15 = 0$

SKILLS CHECK 3D (page 37)

1 c $(0, 4), \sqrt{17}$ **d** $x^2 + (y - 4)^2 = 17$

2 $k = 4, -1$

3 b $x = 8$ **c** $p = 8, q = -1$

4 b Yes. C on perp. bisector of AB

5 a $(1, 2)$

6 a $y = 2x - 2$ **b** $y = -x + 4$ **c** $(2, 2)$

7 $y = \frac{1}{3}x - \frac{10}{3}$

9 $y = -x + 4$

10 $y = \frac{1}{2}x + \frac{3}{2}$

11 a $y = 4x - 3, y = -\frac{1}{4}x - 3$ **b** $(0, -3)$

SKILLS CHECK 3E (page 41)

1 a $(3, 4), (4, 3)$ **b** $(2, 4)$ **c** $(1, 3), (4, 6)$ **d** $(-2, -1)$

2 $x = 1, y = -2$

4 a $x = 1, y = 0$ **b** $x = 0, y = 0; x = 6, y = 0$
  c $x = 2, y = 8$ **d** None

5 $k = 1$

6 b $(-3, 4), (-4, 3)$

7 a $2x + 3y = 17$ **b** $(1, 5)$ **c** $(x - 4)^2 + (y - 3)^2 = 13$

SKILLS CHECK 3F (page 46)

2 a $(1 - 2x)(x + 4)$

4 a $(x + 6)(x - 4), x = -6, x = 4$ **b** $(x + 1)^2 - 25$ **d** $x = -1$

6 b $(0, -2), (-1, 0), (-0.5, 0), (2, 0)$

7 $(4, 16) (0, 0)$

8 $(-\frac{1}{2}, -1)$

SKILLS CHECK 3G (page 52)

1 a i $\begin{pmatrix} 0 \\ 3 \end{pmatrix}$ **ii** $(0, 3), 3$ **b i** $\begin{pmatrix} 0 \\ -2 \end{pmatrix}$ **ii** $(0, -2), -2$

  c i $\begin{pmatrix} 0 \\ 1 \end{pmatrix}$ **ii** $(0, 1), 1$ **d i** $\begin{pmatrix} -2 \\ 0 \end{pmatrix}$ **ii** $(-2, 0), 4$

  e i $\begin{pmatrix} 1 \\ 0 \end{pmatrix}$ **ii** $(1, 0), 1$ **f i** $\begin{pmatrix} -4 \\ 0 \end{pmatrix}$ **ii** $(-4, 0), 16$

  g i $\begin{pmatrix} -3 \\ 1 \end{pmatrix}$ **ii** $(-3, 1), 10$ **h i** $\begin{pmatrix} 2 \\ -1 \end{pmatrix}$ **ii** $(2, -1), 3$

  i i $\begin{pmatrix} -1 \\ -2 \end{pmatrix}$ **ii** $(-1, -2), -1$

2 a i $\begin{pmatrix} -2 \\ 0 \end{pmatrix}$ **ii** $(-2, 0), 2$ **b i** $\begin{pmatrix} 0 \\ 2 \end{pmatrix}$ **ii** $(0, 2), 5$

  c i $\begin{pmatrix} -1 \\ 3 \end{pmatrix}$ **ii** $(-1, 3), 3$ **d i** $\begin{pmatrix} 2 \\ -1 \end{pmatrix}$ **ii** $(2, -1), 2$

3 a $(x - 1)^2 + (y - 2)^2 = 4$ **b** $(x + 1)^2 + (y + 1)^2 = 9$

4 a i Translation $\begin{pmatrix} 0 \\ 2 \end{pmatrix}$ **ii** Reflection in the x-axis

5 a i Stretch in the y-direction, factor 2
  ii Reflection in the x-axis.

6 a Stretch in the y-direction, factor 2
  b Stretch in the x-direction, factor $\frac{1}{2}$
  c Translation $\begin{pmatrix} -1 \\ 0 \end{pmatrix}$

7 b i Stretch in the y-direction, factor 2
  ii Curve goes through $(0, 0), (1, 6), (2, 0); A(1, 6)$

8 Translation $\begin{pmatrix} -1 \\ -4 \end{pmatrix}$

9 a i Stretch in the y-direction, factor 4 **ii** Stretch in the x-direction, factor $\frac{1}{2}$
  b $y = 4x^2$; the curves are the same

10 a Graph through $(-2, 0), (-1, -0.5), (0, 0), (2, 1)$
  b Graph through $(-2, -1), (0, 0), (1, 0.5), (2, 0)$
  c Graph through $(-2, 0), (0, -1), (1, -1.5), (2, -1)$

11 a and **d**, both are $y = 9x^2$

12 a i Reflection in the x-axis
  ii Reflection in the y-axis
  b $y = x^3$, the transformed curves are the same.

13 a Stretch in the y-direction, factor 2; $y = 2\sin x$
  b Translation $\begin{pmatrix} 0 \\ 1 \end{pmatrix}$; $y = 1 + \sin x$
  c Stretch in the x-direction, factor $\frac{1}{2}$; $y = \sin 2x$.

Exam practice 3 (page 54)

1 iii $y(x) = -x^2$ $h(x) = -x^2 + 2$

2 ii $y = 6\sqrt{x}$ **iii** Translation $\begin{pmatrix} k \\ 0 \end{pmatrix}$

3 ii $k < 0, k > \frac{3}{4}$ iii $k = 0, k = \frac{3}{4}$

4 i $x = 3, y = 2$
ii $y = 3x - 7$ is a tangent to $y = x^2 - 3x + 2$ at $(3, 2)$
iii $x + 3y - 9 = 0$

5 i $5, (6, 6)$ iii $(7\frac{3}{4}, 0)$

6 i -71 ii 0

7 i $(3.5, 0.5)$ ii $\dfrac{4 - k}{2}$ iii $k = -2, 3$ iv 15 units^2

8 i $2x + 3y = 9$ iii 52π units^2

9 i $\frac{1}{2}, -2, -\frac{1}{3}$

10 i 1 iii $y - 4 = -(x - 3)$ iv 2 v 12 units^2

11 $(-0.5, 2.5)$

SKILLS CHECK 4A (page 60)

1 a $4x^3$ b $21x^2 - 4x$ c $32x + 8$ d $2x - \frac{1}{2}$
   e $-3x^{-4}$ f $-10x^{-6}$ g $10x^{\frac{2}{3}}$ h $-3x^{-\frac{3}{2}}$

2 a $3x^2 - 4x + 1$ b $6x^3 + \frac{5}{2}$ c 0

3 $-2x^{-3}$

4 a $\frac{1}{4}x^{-\frac{3}{2}}$ b $\frac{1}{4}$

5 a $3x^2 - 6x$ b $(1, -2)$

6 5

7 a $8x + 17$ b $5 - 6y$

8 $\frac{3}{8}, 3, -\frac{1}{9}$

9 ± 2

10 $\pm\frac{1}{2}\sqrt{2}$

11 -4

13 42

SKILLS CHECK 4B (page 63)

1 $y = 3x - 5$

2 a 2 b 5 c $y = 5x - 3$ d $x + 5y - 11 = 0$

3 a i $x > 2$ ii $x < 2$
b i $x > 0$ ii $x < 0$
c i $x > 1, x < -1$ ii $-1 < x < 1$

4 a -2 b $y = \frac{1}{2}x + 3$ c $\frac{5}{2}$

5 a $y = \frac{1}{2}x + \frac{9}{2}$ b $\left(\frac{3}{2}, \frac{21}{4}\right)$

6 a $y = -6x$ b $(1, -2)$

7 $x > \dfrac{1}{12}$

8 a $y = 4x - 12$ b $A(3, 0), B(0, -12)$ c 18 units^2

9 a $3x^{\frac{1}{2}} - 10x^{\frac{3}{2}} + 2$ c $x - 5y - 1 = 0$

10 $x - 2y + 9 = 0$

11 $y = 5x + 80$

12 b $x + 48y - 32 = 0$

SKILLS CHECK 4C (page 68)

1 a $(0, 0)$ min, $\left(-\frac{2}{3}, \frac{4}{27}\right)$ max   b $(1, -2)$ min, $(-1, 2)$ max
   c $(-2, 26)$ max, $(1, -1)$ min

2 $x = -\frac{1}{2}$ max, $x = \frac{2}{3}$ min

3 $500{,}000$ m^2

4 $10{,}000$

5 b $1, 3$ c 16

6 c $360x - 144x^2; 2.5$ d $A = 360x - 84x^2$

7 $\frac{3}{4}$, minimum

8 a 2 b $8\sqrt{2}$

9 c 972 cm^2

Exam practice 4 (page 70)

1 ii $6 - x$ iii 18 cm^2

2 i $3x^2 - 8x - 3$ ii $6x - 8$ iii $x = -\frac{1}{3}$ or 3
iv $x = -\frac{1}{3}$ maximum, $x = 3$ minimum

3 i $y = 6x + 2$ ii $k = 3$ iii $x + 4y - 12 = 0$

4 ii Translation $\begin{pmatrix} -2 \\ 0 \end{pmatrix}$ iii $\dfrac{-1}{x^2}$ iv $-\frac{1}{4}$

5 $15x^4 - x^{-3}$

6 18

7 i $12x^2$ ii $4x^3 + 4x$ iii $\frac{1}{2}x^{-\frac{1}{2}}$

8 i $(-1, 0), (2, -27)$ ii $(-1, 0)$ maximum $(2, -27)$ minimum
iv Goes through $(-1, 0)$ $(0, -7)$ $(2, -27)$ $(3.5, 0)$

9 $6x - 5$

Practice exam paper (page 72)

1 $4\pi r^2$

2 $a - b$

3 $x = 2$ or $x = -1$

4 i $2 - \sqrt{3}$ ii 0.289

5 i $(-1, 1)$ ii $6\sqrt{5}$ iii $x + 2y - 1 = 0$

6 i $(-2, -1)$; radius $= 3$ ii $\sqrt{5}$; inside the circle

7 $-2 < k < 6$

8 iii $\dfrac{3}{2\sqrt{x}}$ iv $3x - 4y + 12 = 0$
v $y = \dfrac{3x}{4}$; $(0, 0)$ and $(16, 12)$

9 i $\frac{1}{6}$ ii $2x - 4$ iii $\dfrac{-6}{t^3}$

10 i $f'(x) = 3x^2 + 12x - 15$; $f''(x) = 6x + 12$
ii $(-5, -50)$ where $f''(x) = -18$; $(1, -8)$ where $f''(x) = 18$
iii $a = -100$ or $a = 8$

SINGLE USER LICENCE AGREEMENT FOR CORE 1 FOR OCR CD-ROM
IMPORTANT: READ CAREFULLY

WARNING: BY OPENING THE PACKAGE YOU AGREE TO BE BOUND BY THE TERMS OF THE LICENCE AGREEMENT BELOW.

This is a legally binding agreement between You (the user or purchaser) and Pearson Education Limited. By retaining this licence, any software media or accompanying written materials or carrying out any of the permitted activities You agree to be bound by the terms of the licence agreement below.

If You do not agree to these terms then promptly return the entire publication (this licence and all software, written materials, packaging and any other components received with it) with Your sales receipt to Your supplier for a full refund.

YOU ARE PERMITTED TO:

- Use (load into temporary memory or permanent storage) a single copy of the software on only one computer at a time. If this computer is linked to a network then the software may only be used in a manner such that it is not accessible to other machines on the network.

- Transfer the software from one computer to another provided that you only use it on one computer at a time.

- Print a single copy of any PDF file from the CD-ROM for the sole use of the user.

YOU MAY NOT:

- Rent or lease the software or any part of the publication.

- Copy any part of the documentation, except where specifically indicated otherwise.

- Make copies of the software, other than for backup purposes.

- Reverse engineer, decompile or disassemble the software.

- Use the software on more than one computer at a time.

- Install the software on any networked computer in a way that could allow access to it from more than one machine on the network.

- Use the software in any way not specified above without the prior written consent of Pearson Education Limited.

- Print off multiple copies of any PDF file.

ONE COPY ONLY

This licence is for a single user copy of the software

PEARSON EDUCATION LIMITED RESERVES THE RIGHT TO TERMINATE THIS LICENCE BY WRITTEN NOTICE AND TO TAKE ACTION TO RECOVER ANY DAMAGES SUFFERED BY PEARSON EDUCATION LIMITED IF YOU BREACH ANY PROVISION OF THIS AGREEMENT.

Pearson Education Limited and/or its licensors own the software.
You only own the disk on which the software is supplied.

Pearson Education Limited warrants that the diskette or CD-ROM on which the software is supplied is free from defects in materials and workmanship under normal use for ninety (90) days from the date You receive it. This warranty is limited to You and is not transferable. Pearson Education Limited does not warrant that the functions of the software meet Your requirements or that the media is compatible with any computer system on which it is used or that the operation of the software will be unlimited or error free.

You assume responsibility for selecting the software to achieve Your intended results and for the installation of, the use of and the results obtained from the software. The entire liability of Pearson Education Limited and its suppliers and your only remedy shall be replacement free of charge of the components that do not meet this warranty.

This limited warranty is void if any damage has resulted from accident, abuse, misapplication, service or modification by someone other than Pearson Education Limited. In no event shall Pearson Education Limited or its suppliers be liable for any damages whatsoever arising out of installation of the software, even if advised of the possibility of such damages. Pearson Education Limited will not be liable for any loss or damage of any nature suffered by any party as a result of reliance upon or reproduction of or any errors in the content of the publication.

Pearson Education Limited does not limit its liability for death or personal injury caused by its negligence.

This licence agreement shall be governed by and interpreted and construed in accordance with English law.